QUANTUM PHYSICS

The Foundations of Mechanics, Entanglement and Wave Theory

A Beginners Guide to Quantum Mechanics, String Theory, Field Theory and Quantum Computing

By Isaac Archer

© Copyright 2024 Isaac Archer - All rights reserved.

The content within this book may not be reproduced, duplicated or transmitted without direct written permission from the author or the publisher.

Under no circumstances will any blame or legal responsibility be held against the publisher, or author, for any damages, reparation, or monetary loss due to the information contained within this book, either directly or indirectly. You are responsible for your own choices, actions and results.

Legal Notice:

This book is copyright protected. This book is only for personal use. You cannot amend, distribute, sell, use, quote or paraphrase any part, of the content within this book, without the consent of the author or publisher.

Disclaimer Notice:

Please note the information contained within this document is for educational and entertainment purposes only. All effort has been expended to present accurate, up-to-date, and reliable, complete information. No warranties of any kind are declared or implied. Readers acknowledge that the author is not engaging in the rendering of legal, financial, medical or professional advice. The content within this book has been derived from various sources. Please consult a licensed professional before attempting any techniques outlined in this book.

By reading this document, the reader agrees that under no circumstances is the author responsible for any losses, direct or indirect, which are incurred as a result of the use of the information contained within this document, including, but not limited to errors, omissions or inaccuracies.

Quantum Physics:
The Foundations of Mechanics, Entanglement and Wave Theory
Copyright © 2024 Isaac Archer
ISBN: 979-8-3249-7403-9

List of Contents

Introduction 9

Chapter 1: The Quantum World Unveiled 12
1.1 The Story Behind Quantum Mechanics - A Historical Overview 12
1.2 Wave-Particle Duality 15
1.3 The Uncertainty Principle Demystified 17
1.4 Quantum States and Superposition 19
1.5 Quantum Entanglement 21
1.6 The Role of Observers in Quantum Mechanics - Collapsing the Wave Function 24

Chapter 2: The Mathematical Symphony of Quantum Mechanics 30
2.1: Linear Algebra in Quantum Mechanics 30
2.2: Complex Numbers and Their Role in Quantum Physics 33
2.3: Differential Equations as the Language of Quantum Mechanics 35
2.4: Operators and Eigenvalues - A Beginner's Guide 37
2.5: The Schrödinger Equation 39
2.6: Dirac Notation 42

Chapter 3: Beyond Classical Boundaries 48
3.1: The Principle of Superposition and Its Applications in Quantum Computing 48
3.2: Quantum Tunneling in Semiconductors and Technological Impacts 50
3.3: Quantum Entanglement in Cryptography 52
3.4: The Quantum Zeno Effect 54
3.5: Wave Function Collapse 56
3.6: Decoherence 58

Chapter 4: Quantum Mechanics at Work 62
4.1: Solving the Schrödinger Equation 62
4.2: Quantum Mechanics in Chemistry 64
4.3: Quantum Computing 66
4.4: Quantum Mechanics in Everyday 68
4.5: Practical Quantum Experiments You Can Try at Home 70
4.6: The Future of Quantum Technologies 72

Chapter 5: Bridging Quantum Mechanics and Relativity 78
5.1: Special Relativity and Quantum Mechanics 78
5.2: The Klein-Gordon Equation 81
5.3: The Dirac Equation 83
5.4: Quantum Field Theory - Basics for Beginners 85
5.5: The Standard Model of Particle Physics Simplified 88
5.6: Gravitational Quantum Mechanics 90

Chapter 6: Quantum Paradoxes and Thought Experiments 96

6.1: Schrödinger's Cat and the Observer's Role 96
6.2: The EPR Paradox - Questioning Locality and Reality 99
6.3: The Many-Worlds Interpretation vs. Copenhagen Interpretation 102
6.4: Quantum Eraser Experiment 106
6.5: Wigner's Friend 109
6.6: The Delayed Choice Quantum Eraser 112

Chapter 7: Quantum Mechanics for the Curious Mind 118
7.1: Addressing Common Misconceptions in Quantum Mechanics 118
7.2: Quantum Mechanics in Popular 121
7.3: How to Think Like a Quantum Physicist 124
7.4: The Future of Quantum Research 127
7.5: Quantum Mechanics and Consciousness 131
7.6: Learning Resources and How to Further Your Quantum Journey 135

Chapter 8: Quantum Mechanics Demystified - A Practical Guide 144
8.1: Step-by-Step Guide to Solving Quantum Problems 144
8.2: Case Studies in Modern Technologies 150
8.3: Quantum Simulation Software 155
8.4: Developing Intuition for Quantum Phenomena through Visualization 160
8.5: Quantum Mechanics in Action 165
8.6: Preparing for Advanced Study in Quantum Physics 169

A Quantum Leap Forward 175

Introduction

Once upon a time, there was a student much like many of you - filled with curiosity and ambition, yet standing at the edge of the quantum realm, both enthralled and slightly overwhelmed. The student's journey into quantum mechanics began with a simple question about the nature of light. Still, it quickly unfolded into a saga of particles acting as waves, cats being simultaneously alive and dead, and probabilities ruling the universe. This story is not unique; it mirrors the experience of countless individuals who have encountered quantum physics' mysterious yet captivating world.

This book's purpose is straightforward: to untangle the intricate web of quantum mechanics for students, beginners and non-specialists. It is crafted to be your compass through the quantum world, guiding you from the shores of classical physics to the quantum depths with clarity and confidence.

Quantum physics often appears cloaked in a shroud of abstract concepts and daunting mathematics, making it seem impenetrable to the uninitiated. This book acknowledges those challenges head-on, striving to clear the fog and demystify the subject. It is tailored for those who have felt lost in the jargon or bogged down by equations, providing a lifeline to grasp the beauty and elegance of quantum physics without getting lost in the complexities.

Within these pages, you'll find a carefully structured exploration of quantum mechanics, from its foundational principles to its mathematical tools—presented in a simplified manner—and onward to its astonishing applications. The book progresses logically, ensuring you build a solid foundation before venturing into more nuanced topics. This step-by-step approach fosters a deeper understanding and appreciation of the subject.

As the author of this book, my journey into the quantum world was driven by a deep-seated passion and endless fascination with how the universe operates at its most fundamental level. I write to share this passion with you, making quantum mechanics accessible to a broader audience.

Approach this book with an open mind and a readiness to embrace new concepts. I encourage you to dive into this book with the same curiosity and eagerness that has propelled scientists and thinkers for over a century.

As we stand on the brink of discovery, let us ponder: "If quantum mechanics hasn't profoundly shocked you, you haven't understood it yet." This sentiment captures the essence of the quantum world—a place where our intuition is challenged, and our curiosity is the key to unlocking profound truths. Are you ready to explore the depths of quantum mechanics and see the universe in a new light?

1

The Quantum World Unveiled

Chapter 1: The Quantum World Unveiled

Welcome to the fascinating world of quantum mechanics! In this chapter, we'll embark on a captivating journey through the history, key concepts, and mind-bending implications of this revolutionary field of physics. Prepare to explore the realm of the incredibly small, where the rules of classical physics no longer apply, and the behavior of particles defies our everyday intuition.

1.1 The Story Behind Quantum Mechanics - A Historical Overview

Picture this: it's the early 20th century, and physicists are grappling with some perplexing problems that classical physics can't seem to solve. The world of the very small, the domain of atoms and subatomic particles, follows a different set of rules than the macroscopic world we're used to. That's when a group of brilliant minds, including Albert Einstein, Max Planck, Niels Bohr and Werner Heisenberg, began to unravel the mysteries of the subatomic world, laying the foundation for what we now know as quantum mechanics.

Pioneers of Quantum Theory

• Max Planck kicked things off in 1900 by proposing that energy comes in discrete packets called "quanta," introducing the famous Planck constant. This groundbreaking idea marks the birth of quantum physics and sets the stage for future revolutionary developments.

• In 1905, Albert Einstein took it a step further, suggesting that light is quantized and made up of particles called photons. This groundbreaking idea explains the photoelectric effect, where light can eject electrons from a metal surface, and challenges the classical wave theory of light. Einstein's work on the photoelectric effect earned him the Nobel Prize in Physics in 1921.

• Niels Bohr, building on Ernest Rutherford's model of the atom, introduced the Bohr model in 1913, incorporating quantum theory to explain the stability of atomic orbits and the emission of light. Bohr's model successfully accounts for the discrete energy levels in atoms and the spectral lines they emit, marking a significant breakthrough in our understanding of atomic structure.

• In 1925, Werner Heisenberg developed matrix mechanics and the uncertainty principle, which states that specific pairs of physical properties, like position and momentum, can never be precisely measured simultaneously. Heisenberg's work lays the foundation for the mathematical formalism of quantum mechanics and highlights the inherent limitations in our ability to observe the quantum world.

The Birth of a New Physics

As these brilliant minds delved deeper into the subatomic realm, it became clear that classical physics couldn't keep up. The behavior of particles at the subatomic and atomic scales defied the predictable, deterministic nature of classical mechanics. Phenomena such as the blackbody radiation problem, the photoelectric effect, and the stability of atomic orbits couldn't be fully explained using classical physics tools.

This realization marked the birth of quantum mechanics, a new physics that embraced the probabilistic nature of the quantum world. The field of quantum mechanics has revolutionized the world of physics, introducing novel concepts that have challenged our classical intuition and reshaped our understanding of the fundamental nature of reality. Concepts like wave-particle duality, superposition, and entanglement have opened up new avenues for research and exploration in the realm of quantum mechanics.

Key Experiments

Two groundbreaking experiments played a crucial role in revealing the quantum nature of reality:

1. The double-slit experiment demonstrated the wave-particle duality of light and matter. This experiment showed that particles, like electrons, can behave like waves, producing interference patterns when passed through two slits. The results of this experiment couldn't be explained by classical physics, and the need for a new quantum framework was highlighted.

2. The photoelectric effect, explained by Einstein, proved that light behaves as particles (photons) and can instantly eject electrons from metal surfaces, challenging the classical wave theory of light. This experiment provided strong evidence for the quantization of light and the particle-like behavior of photons.

These experiments and others, like the Compton effect and the Stern-Gerlach experiment, provided the empirical foundation for developing quantum mechanics. They revealed the limitations of classical physics and paved the way for a new understanding of the subatomic world.

Philosophical Implications

As quantum mechanics took shape, it challenged our fundamental understanding of reality. Concepts like superposition (particles existing in multiple states simultaneously) and entanglement (particles remaining connected despite vast distances) defied classical intuition. They sparked intense philosophical debates about the nature of reality, the role of the observer, and the limits of human knowledge.

The probabilistic nature of quantum mechanics, encapsulated in the wave function, raised questions about determinism and free will. If the quantum world is governed by probability, does this mean that the future is not entirely predictable from the present? The measurement problem arising from the apparent collapse of the wave function upon observation led to various interpretations of quantum mechanics, each with its philosophical implications.

Quantum mechanics has long been a topic of discussion among scientists and philosophers. The profound implications of this field of study on our understanding of the universe and our place within it have been a subject of debate that continues to this day. Many philosophical questions have arisen as scientists and philosophers grapple with the complex nature of quantum mechanics.

1.2 Wave-Particle Duality

The concept of wave-particle duality is one of quantum mechanics' most fascinating and mind-bending aspects. In the quantum world, light and matter can behave as waves and particles, depending on how we observe them. This dual nature challenges our classical intuition and requires a new way of thinking about the universe's fundamental building blocks.

Light as a Wave and Particle

The double-slit experiment is a perfect illustration of the wave-particle duality of light. When light passes through a pair of closely spaced slits and falls on a screen, it creates an interference pattern, just like waves in water. This interference pattern suggests that light is behaving as a wave, with peaks and troughs that can interfere with each other.

However, when we detect individual photons, the fundamental particles of light, they behave as particles, seeming to pass through one slit or the other. This particle-like behavior is evident in the photoelectric effect, where photons can eject electrons from a metal surface.

Classical physics cannot explain light's dual nature as a wave and a particle. It requires a quantum mechanical description, where light is understood as a quantum object that exhibits wave-like and particle-like properties, depending on the experimental context.

Matter Waves

In 1924, French physicist Louis de Broglie suggested that particles might also behave like waves if light can behave like particles. This revolutionary idea, known as the de Broglie hypothesis, suggested that every particle has a wavelength associated with it, inversely proportional to its momentum.

De Broglie's hypothesis was later confirmed by experiments showing that electrons and even larger particles like atoms and molecules can produce interference patterns, just like light waves. This discovery showed that the wave-particle duality is not limited to light but is a fundamental feature of all quantum entities.

Implications for Physics

The wave-particle duality of light and matter required a complete rethinking of the fundamental nature of reality. Classical physics, with its clear distinction between waves and particles, could no longer adequately describe the behavior of quantum entities. This realization led to the development of a new framework, quantum mechanics, which embraces the probabilistic nature of the subatomic world.

Quantum mechanics introduces the wave function, a mathematical object that describes a system's quantum state. The wave function encapsulates quantum entities' probabilistic nature, providing a way to calculate the likelihood of observing a particle in a particular state or location.

The Schrödinger equation is a key quantum mechanics equation describing how a system's quantum state evolves over time. The equation is a wave equation that reflects the wave-like characteristics of quantum particles. It is a crucial component of comprehending how these entities behave on a quantum level.

The Concept of Quantum Objects

In the quantum realm, objects don't neatly fit into the classical categories of "particle" or "wave." Instead, they exhibit both properties, depending on how we measure them. This dual nature is a fundamental feature of quantum objects, challenging our everyday intuition and requiring a new way of thinking about the universe's building blocks.

These properties are only sometimes well-defined, and their values are subject to the laws of quantum mechanics, such as the uncertainty and superposition principles.

The behavior of quantum objects is inherently probabilistic, meaning we can only predict the likelihood of observing a particular outcome, rather than determining it with certainty. This probabilistic nature is a fundamental aspect of quantum mechanics and reflects the inherent randomness and unpredictability of the quantum world.

1.3 The Uncertainty Principle Demystified

The uncertainty principle is one of quantum mechanics' most profound and perplexing concepts. Formulated by Werner Heisenberg in 1927, this principle states that there's a fundamental limit to the precision with which specific pairs of physical properties can be simultaneously known. The uncertainty principle is an essential principle of quantum mechanics, which has significant implications for comprehending the subatomic realm.

Heisenberg's Principle

If you want to determine the position and momentum of a particle at the same time, you would be limited by the uncertainty principle. This principle states that the more accurately you measure one of these properties, the less accurately you can measure the other one. This is not just a limitation of our measuring instruments, but it is a fundamental characteristic of the quantum world.

Mathematically, the uncertainty principle states that the product of the uncertainties in position and momentum must be equal or greater than a particular value, proportional to Planck's constant. This means that as the uncertainty in position decreases, the uncertainty in momentum must increase, and vice versa.

The uncertainty principle applies to position and momentum, as well as other pairs of complementary variables, such as energy and time. The more precisely we know a system's energy, the less precisely we can know the time at which that energy was measured, and vice versa.

Underlying Science

The uncertainty principle arises from the wave-particle duality of quantum entities. Since particles can behave like waves, pinpointing their exact location becomes tricky. The more you try to localize a particle, the more you disturb its momentum, and vice versa. This inherent trade-off is mathematically enshrined in Heisenberg's inequality.

The uncertainty principle is a direct result of the quantum objects' wave nature. In classical physics, a particle can have a well-defined position and momentum simultaneously. However, in quantum mechanics, a particle is described by a wave function that extends over space. The wave function indicates where a particle is likely to be found, with its wavelength corresponding to its momentum.

We must use a wave function with a very short wavelength corresponding to high momentum to localize a particle precisely. Conversely, we need a wave function with a long wavelength to measure the momentum precisely, which spreads out over space and makes it difficult to determine the particle's exact position.

Impact on Observation

Observing a quantum system inevitably disturbs a particle. For example, when you measure a particle's position, you interact with it in a way that alters its momentum. This interaction is not a mere side effect but a fundamental consequence of the quantum nature of reality.

In quantum mechanics, measurement is an active process that collapses the wave function, forcing the quantum system into a definite state. Before the measurement, the system exists in a superposition of multiple states, each with its own probability. When measured, the wave function collapses and the system is found in one state with a probability determined by quantum mechanics.

This wave function collapse upon measurement is a crucial aspect of the quantum measurement problem. It has led to many quantum mechanics interpretations, such as the Copenhagen and many-worlds interpretation and the pilot wave theory.

Philosophical and Practical Consequences

The uncertainty principle has significant implications for the way we understand the universe. It challenges the deterministic worldview of classical physics, suggesting that the future is not entirely predictable from the present. This indeterminacy has sparked philosophical debates about the nature of reality, free will, and the limits of human knowledge.

If the quantum world is governed by inherent uncertainty, does this mean the universe is fundamentally random? Does the uncertainty principle imply that there are limits to what we can know about the world? Physicists and philosophers have long debated the implications of quantum mechanics, and they continue to shape our understanding of quantum mechanics' implications.

The uncertainty principle also has practical consequences in fields like quantum cryptography. By harnessing the inherent uncertainty of quantum systems, researchers have developed unbreakable encryption methods that rely on quantum mechanics laws to ensure security. In quantum cryptography, any attempt to intercept or measure the quantum key without authorization would inevitably disturb the system, revealing the presence of an eavesdropper.

Moreover, the uncertainty principle limits the precision of measurements and technology miniaturization. As we push the boundaries of nanoscale engineering and quantum computing, the uncertainty principle reminds us that there are inherent limitations to how precisely we can control and manipulate matter at the quantum level.

1.4 Quantum States and Superposition

In quantum mechanics, particles can exist in multiple states simultaneously, known as superposition. This concept is mind-boggling and goes against our everyday intuition. It also creates possibilities that cannot be explained using classical logic. Superposition is a crucial aspect of quantum systems, and it has significant implications for our comprehension of the nature of reality.

Quantum States Explained

A quantum state is a mathematical description of a quantum system's properties, such as position, momentum, energy and spin. Unlike classical states with well-defined values, quantum states are probabilistic, representing the likelihood of finding a particle in a particular configuration.

In quantum mechanics, the state of a system is defined by a wave function, which is represented by the Greek letter psi (ψ). The wave function is a function of complex numbers that contains all the necessary information about the quantum system. By squaring the absolute value of the wave function, we can determine the probability of finding the particle at a specific location or with certain properties.

The Schrödinger equation is an essential equation of quantum mechanics that governs the evolution of a system's wave function over time. The Schrödinger equation is deterministic, meaning that the equation can predict the system's future state given an initial quantum state.

Superposition Principle

According to the superposition principle, a quantum system can exist in multiple states simultaneously, each with its own probability. It's only when we measure the system that it "collapses" into a single state. This means that a particle can be in two (or more) places simultaneously or have multiple property values simultaneously.

Mathematically, the superposition principle states that if a quantum system can be in state A or state B, then it can also be in a linear combination of those states, written as $\alpha|A\rangle + \beta|B\rangle$, where α and β are complex numbers that determine the probabilities of finding the system in each state.

The superposition of states is a fundamental feature of quantum systems. It is responsible for many counterintuitive phenomena, such as the double-slit experiment and quantum entanglement, observed in the quantum world.

Famous Thought Experiments

One of the most famous illustrations of superposition is Schrödinger's cat. In 1935, physicist Erwin Schrödinger proposed a thought experiment involving a cat sealed in a box with a device that may or may not kill it based on the state of a quantum particle.

The quantum particle is in a superposition of two states: one that triggers the device and kills the cat and another that leaves the cat alive. According to quantum mechanics, until the box is opened and the cat's state is observed, the cat is simultaneously alive and dead, existing in a superposition of both states.

Schrödinger's cat thought experiment highlights the bizarre consequences of applying quantum mechanics to macroscopic objects and challenges our intuition about the nature of reality. It illustrates the importance of measurement in quantum mechanics and the observer's role in collapsing the wave function.

Real-World Applications

Superposition isn't just a theoretical concept; it has real-world applications in fields like quantum computing and quantum cryptography. In quantum computing, the power of superposition is harnessed to perform complex calculations that would be infeasible for classical computers.

Quantum computers use qubits, or quantum bits, which can exist simultaneously in a superposition of multiple states. By manipulating these qubits and exploiting the principles of superposition and entanglement, quantum computers can solve specific problems exponentially faster than classical computers, such as factoring large numbers and simulating complex quantum systems.

Superposition is utilized in quantum cryptography to establish safe communication channels that are impervious to eavesdropping. Quantum cryptography encodes information in photons' quantum states and sends them via a communication channel. Attempts to measure or intercept the photons will inevitably disturb their state, thus revealing an eavesdropper.

Moreover, superposition plays a crucial role in quantum sensing and metrology, enabling the development of ultra-sensitive detectors and precision measurement devices. By harnessing the sensitivity of quantum systems to external perturbations, researchers can create sensors that can detect tiny changes in magnetic fields, gravitational fields, and other physical quantities with unprecedented accuracy.

1.5 Quantum Entanglement

Entanglement is one of the most bizarre and fascinating phenomena. Entanglement occurs when two or more particles become so inextricably linked that their quantum states cannot be described independently, no matter how far apart they may be. Albert Einstein famously referred to it as "spooky action at a distance," which challenges our understanding of locality and has significant implications for our perception of reality.

Einstein, Podolsky, and Rosen (EPR) Paradox

In 1935, Einstein, Boris Podolsky and Nathan Rosen published a seminal paper that challenged the completeness of quantum mechanics. The EPR paper proposed a thought experiment involving two entangled particles separated by a big distance. By measuring the position or momentum of one particle, the position or momentum of the other particle could be instantaneously determined without any apparent interaction between the particles.

Einstein argued that entanglement violates the principle of locality, which states that objects can only be influenced by their immediate surroundings. He believed that quantum mechanics must be incomplete and that there must be hidden variables that determine the outcomes of measurements on entangled particles.

The Nature of Entangled States

When particles become entangled, their quantum states are correlated so that measuring one particle instantly determines the state of the other, regardless of the distance between them. This correlation does not result from any classical interaction or communication between the particles but is a fundamental feature of the quantum world.

Entanglement arises from the superposition principle of quantum mechanics. When two particles interact, their wave functions can combine to create a single, inseparable quantum state. This entangled state cannot be described as a product of individual particle states but must be treated as a whole.

The correlations between entangled particles are stronger than any classical correlations and cannot be explained by any local hidden variable theory. Entanglement is a quantum phenomenon with no classical analog, challenging our intuition about the nature of reality.

Bell's Theorem and Experimental Verification

In 1964, physicist John Stewart Bell proved a groundbreaking theorem that provided a way to test the EPR paradox experimentally. Bell's theorem showed that any local hidden variable theory that aimed to explain the correlations between entangled particles must satisfy certain statistical limits, known as Bell's inequalities.

On the other hand, quantum mechanics predicts that Bell's inequalities can be violated under certain conditions. If experiments could demonstrate a violation of Bell's inequalities, it would prove that any local hidden variable theory cannot explain the correlations between entangled particles and that quantum mechanics is a complete description of reality.

Since the 1970s, numerous experiments have been conducted to test Bell's inequalities using entangled photons, electrons and even atoms. These experiments have consistently shown a violation of Bell's inequalities, providing strong evidence for the reality of quantum entanglement and the validity of quantum mechanics.

The most famous of these experiments is the Aspect experiment, conducted by French physicist Alain Aspect and his team in 1982. The Aspect experiment used a source of entangled photons and two detectors separated by a considerable distance. By measuring the polarization of the photons at each detector, the team demonstrated a clear violation of Bell's inequalities, confirming the predictions of quantum mechanics.

Implications for Information Theory and Cryptography

Entanglement has a far-reaching impact on information theory and cryptography. It enables the development of quantum cryptography protocols, such as quantum key distribution, which uses entangled particles to create unbreakable encryption keys.

In quantum key distribution, a sender and a receiver share a pair of entangled particles and use them to generate a secret key. By measuring the quantum states of the particles, the sender and receiver can create a key that is perfectly correlated and secure from eavesdropping. Attempts to intercept or measure the particles would inevitably disturb their entangled state, revealing the presence of an eavesdropper.

Moreover, entanglement is a crucial resource for quantum computing and information processing. Entangled qubits perform complex calculations in quantum computing and simulate quantum systems intractable for classical computers. Entanglement allows quantum computers to explore vast computational spaces simultaneously, enabling them to solve problems exponentially faster than classical computers.

Entanglement also plays a crucial role in quantum error correction, essential for building large-scale, fault-tolerant quantum computers. By distributing information across multiple entangled qubits, quantum error correction schemes can detect and correct errors arising from decoherence and other noise sources, ensuring the reliability and scalability of quantum computations.

1.6 The Role of Observers in Quantum Mechanics - Collapsing the Wave Function

In quantum mechanics, the observer plays a crucial role in determining the outcome of measurements. This is exemplified by wave function collapse, which describes how a quantum system's superposition of states is reduced to a single state upon observation. The role of the observer has been a subject of intense debate and has led to various interpretations of quantum mechanics, each with its own philosophical implications.

Observation in Quantum Mechanics

In the quantum world, observation is not a passive process. When we measure a quantum system, we interact with it in a way that fundamentally alters its state. This interaction causes the system's wave function, representing its superposition of states, to collapse into a single state.

The wave function, represented by the Greek letter psi (ψ), is a mathematical entity that encapsulates all the information about a quantum system. It is a complex-valued function that expresses the probability amplitude of locating the system in a specific state. The square of the absolute value of the wave function provides the probability density of observing the system in that state.

Before measurement, the quantum system exists in a superposition of multiple states, each with its probability amplitude. Based on the Schrödinger equation, the wave function changes over time in a deterministic way, predicting the system's future state based on its initial conditions.

Wave Function Collapse

When an observer measures a quantum system, the wave function collapses, and the system is found in a single state. This process is instantaneous and irreversible, one of the most puzzling aspects of quantum mechanics.

The collapse of the wave function is a probabilistic process governed by the Born rule. The Born rule states that the probability of observing a particular outcome is given by the square of the absolute value of the probability amplitude associated with that outcome.

For example, if a quantum system is in a superposition of two states, $|A\rangle$ and $|B\rangle$, with probability amplitudes α and β, respectively, then the probability of observing the system in state $|A\rangle$ upon measurement is $|\alpha|^2$, and the probability of observing it in state $|B\rangle$ is $|\beta|^2$.

The collapse of the wave function is a controversial and poorly understood aspect of quantum mechanics. It raises fundamental questions about the nature of reality and the role of the observer in determining the outcome of measurements.

Debates and Interpretations

The observer's role in quantum mechanics has led to various interpretations of the theory and sparked vigorous debates. These interpretations attempt to reconcile quantum mechanics' counterintuitive aspects with our classical intuition and provide a coherent framework for understanding the nature of reality.

The Copenhagen interpretation, developed by Niels Bohr and Werner Heisenberg, is one of the most widely accepted interpretations of quantum mechanics. In this particular interpretation, the wave function is not a physical object but rather a mathematical tool that represents our understanding of a quantum system. When a measurement is taken, the wave function collapses, and the outcome of the measurement is determined by probability according to the Born rule.

The Many-Worlds interpretation, proposed by Hugh Everett in 1957, takes a radically different approach. According to this interpretation, the wave function never collapses. Instead, whenever a measurement is made, the universe splits into multiple parallel branches, each representing a different measurement outcome. In this view, all possible measurement outcomes co-occur in different parallel universes.

Other interpretations, such as the de Broglie-Bohm pilot wave theory and the objective collapse theories, attempt to provide alternative explanations for the wave function's collapse and the observer's role in quantum mechanics.

Experimental Evidence

Experiments have demonstrated the crucial role of observation in quantum mechanics. One famous example is the double-slit experiment, which shows the wave-particle duality of matter and the effect of observation on the behavior of quantum systems.

In the double-slit experiment, a beam of particles, such as photons or electrons, is directed towards a screen with two parallel slits. If the particles are not observed, they exhibit an interference pattern on the screen, indicating that they behave like waves and pass through both slits simultaneously.

However, if a detector is placed near the slits to determine which slit each particle passes through, the interference pattern disappears, and the particles behave like classical particles, passing through only one slit at a time.

This experiment demonstrates that observation can fundamentally change a quantum system's behaviour, causing it to switch from a wave-like state to a particle-like state.

Another example is the quantum Zeno effect, where frequent measurements can prevent a quantum system from evolving. By repeatedly measuring a quantum system in short intervals, the system is effectively "frozen" in its initial state, and its evolution is suppressed.

These experiments highlight the intimate connection between observation and reality in the quantum realm and provide evidence for the collapse of the wave function upon measurement.

The role of the observer is also central to quantum computing and cryptography. In quantum computing, measurements extract information from quantum systems and perform computations. The choice of measurement basis and the timing of measurements can significantly impact the outcome of quantum computation.

In quantum cryptography, the observer effect is harnessed to detect eavesdropping attempts. Any unauthorized measurement of the quantum states used to transmit information would inevitably disturb the system, revealing the presence of an eavesdropper and ensuring the security of the communication.

Conclusion

In this chapter, we have explored the fascinating and mind-bending world of quantum mechanics. From its historical development to key concepts like wave-particle duality, the uncertainty principle, superposition, entanglement, and the observer's role, quantum theory challenges our classical intuition and reshapes our understanding of reality.

Quantum mechanics has transformed our comprehension of the subatomic realm, leading to groundbreaking technologies and new fields like quantum computing, cryptography, and sensing. It also raises profound philosophical questions about the nature of reality and the limits of knowledge that continue to be debated.

As we further unravel the mysteries of the quantum world, it is crucial to keep an open mind and embrace its counterintuitive aspects. Doing so can unlock new insights into the fundamental workings of nature, push the boundaries of human knowledge, and remind us of the incredible complexity and beauty of our universe. With each quantum discovery and application, we move closer to a deeper understanding of reality.

The Mathematical Symphony of Quantum Mechanics

Chapter 2: The Mathematical Symphony of Quantum Mechanics

The previous chapter explored this revolutionary theory's historical development, key concepts and mind-bending implications. Now, it's time to explore the mathematical framework that underpins the quantum realm.

But fear not! While the mathematics of quantum mechanics may seem daunting at first, we'll navigate this complex landscape together, step by step. Think of it as learning a new language—the language of the universe at its most fundamental level. Like any language, it has syntax, grammar, and vocabulary, which we'll gradually master together.

In this chapter, we'll explore the essential mathematical tools that form the backbone of quantum theory. From the elegant simplicity of linear algebra to the mysterious beauty of complex numbers, from the dynamic flow of differential equations to the abstract power of operators and eigenvalues, we'll unravel the mathematical tapestry that weaves together the principles of quantum mechanics.

2.1: Linear Algebra in Quantum Mechanics

Basics of Linear Algebra

Linear algebra, a branch of mathematics that deals with transformations of vectors and matrices, lies at the heart of quantum mechanics. You might wonder: "What does linear algebra have to do with the subatomic world?". Well, linear algebra provides the perfect language to describe the states and properties of quantum systems.

Let's use an analogy to understand how linear algebra fits into quantum mechanics. Imagine you're exploring a vast city, with streets and avenues crisscrossing like a grid. Each intersection represents a unique location, and to get from one place to another, you navigate along these streets and avenues, following a specific path. In linear algebra, we call these paths "vectors."

Vectors are mathematical objects that have both direction and magnitude. They can represent physical quantities like position, momentum, or even the state of a quantum system. Just as you can combine different paths to reach your destination in a city, you can add and subtract vectors to describe the combined effect of multiple quantities or transformations.

State Vectors and Hilbert Space

Now, let's take our city analogy a step further. Imagine that each intersection not only represents a location but also a possible state of a quantum system. In quantum mechanics, we call this abstract space "Hilbert space," named after the mathematician David Hilbert.

In Hilbert space, each possible state of a quantum system is represented by a "state vector," often denoted as $|\psi\rangle$ in Dirac notation (which we'll explore later in this chapter). These state vectors encapsulate all the information about the system's properties, such as its position, momentum, or energy.

Just as a GPS helps you navigate a city by providing coordinates, state vectors help us navigate the quantum world by giving the "coordinates" of a quantum system in Hilbert space. The state vector $|\psi\rangle$ is like a quantum GPS, guiding us through the vast landscape of possible states.

Matrix Mechanics

But how do we manipulate these state vectors and perform operations on them? This is where matrices come into play. In quantum mechanics, physical quantities and transformations are represented by matrices, which act on state vectors to produce new states or extract information about the system.

Imagine a city map that shows you how different paths connect and transform into one another. In linear algebra, matrices play a similar role, mapping vectors from one space to another or changing them in specific ways. For example, a rotation matrix can rotate a vector by a certain angle, just like turning a corner in a city.

In quantum mechanics, matrices represent "operators" corresponding to physical observables, such as position, momentum, or energy. When an operator acts on a state vector, it extracts the corresponding physical property of the system. This is like using a particular city map filter to highlight specific features, such as restaurants or parks.

Eigenvalues and Eigenvectors

Now, let's discuss a fundamental concept in linear algebra: eigenvalues and eigenvectors. In our city analogy, imagine a special kind of path that, when followed, always leads back to itself, only scaled by a particular factor. These paths are called "eigenvectors," and the scaling factors are called "eigenvalues."

In quantum mechanics, eigenvectors and eigenvalues play a crucial role in understanding a system's measurable properties. When an operator acts on an eigenvector, the result is the same eigenvector scaled by its corresponding eigenvalue. The eigenvectors of an operator represent the states in which the system has a definite value for the corresponding physical observable.

Think of eigenvalues as the "quantum price tags" attached to each eigenvector. Just as a price tag tells you the cost of an item in a store, an eigenvalue tells you the value of a physical observable when the system is in the corresponding eigenvector state.

2.2: Complex Numbers and Their Role in Quantum Physics

Introduction to Complex Numbers

Now, let's take a moment to appreciate the unsung heroes of quantum mechanics: complex numbers. These mathematical marvels, often overlooked in our everyday lives, play a crucial role in describing the behavior of quantum systems.

At first glance, complex numbers seem like a strange concept. They consist of a natural part and an imaginary part, denoted as a + bi, where a and b are real numbers, and i is the square root of -1. Yes, you read that right: the square root of a negative number! But don't let that intimidate you; complex numbers are a powerful tool for understanding the quantum world.

Complex Numbers in Quantum Mechanics

In quantum mechanics, complex numbers describe the probability amplitudes of quantum states. These amplitudes, represented by the wave function ψ, are complex-valued functions that contain information about the likelihood of finding a particle in a particular state or position.

But why complex numbers? It turns out that using complex numbers is not just a mathematical trick but a fundamental requirement for describing the behavior of quantum systems. The wave-like nature of particles and the phenomenon of interference can only be fully captured by using complex numbers.

Imagine a quantum particle as a tiny surfer riding a complex wave. The real part of the wave function represents the wave's amplitude, while the imaginary part captures its phase. Just as the height and shape of a water wave determine how it interacts with other waves, the amplitude and phase of the quantum wave function determine how the particle behaves and interacts with its environment.

Phase and Interference

The phenomenon of interference is one of the most striking features of quantum mechanics, where particles can exhibit wave-like behavior and interfere with themselves. This feature is beautifully demonstrated in the famous double-slit experiment, where a single particle can pass through two slits simultaneously and create an interference pattern on a screen.

The key to understanding interference lies in the phase of the wave function, which is encoded in the complex nature of probability amplitudes. The phase determines how different parts of the wave function interact with each other, leading to constructive or destructive interference.

Picture two quantum waves meeting at a point in space. If their phases align, they create a more significant wave—this is constructive interference. But when their phases are not aligned, they can cancel each other out due to destructive interference. It's like two dancers moving in perfect synchrony or entirely out of step, their movements either amplifying or nullifying each other.

Real-World Examples

Using complex numbers in quantum mechanics isn't just an abstract concept; it has real-world implications and applications. One fascinating example is the behavior of superconductors, materials that conduct electricity with no resistance at low temperatures.

The theory of superconductivity relies on the quantum mechanical description of electron pairs, known as Cooper pairs. These pairs form a coherent quantum state described by a complex wave function, and it's the phase coherence of this wave function gives rise to the remarkable properties of superconductors.

Imagine a synchronized swimming team, where each pair of swimmers performs a perfectly coordinated routine. In a superconductor, the electron pairs move in unison, their complex wave functions aligned in phase, allowing them to flow through the material without resistance. It's like a quantum ballet, with the electrons dancing to the tune of complex numbers.

2.3: Differential Equations as the Language of Quantum Mechanics

Understanding Differential Equations

Now, let's dive into the heart of quantum mechanics: differential equations. These mathematical statements, which relate a function to its derivatives, are the key to unlocking the secrets of the subatomic world.

In classical physics, differential equations describe the motion of objects, from the trajectory of a thrown ball to the planet's orbit. Differential equations take on a new role in quantum mechanics: they describe the evolution of quantum states over time and space.

Picture a quantum state as a complex landscape, with peaks and valleys representing the probability of finding a particle at different locations. Differential equations act as the "terrain engineers" of this landscape, shaping and molding it as time passes. They tell us how the peaks and valleys shift, grow, or diminish, guiding the evolution of the quantum system.

The Schrödinger Equation

The Schrödinger equation, the cornerstone of quantum mechanics, is at the centre of this mathematical framework. Developed by Austrian physicist Erwin Schrödinger in 1925, this equation describes how the wave function of a quantum system evolves over time.

The Schrödinger equation comes in two flavors: the time-dependent form and the time-independent form. The time-dependent Schrödinger equation describes how the wave function changes with time, capturing the dynamic behavior of quantum systems. The time-independent Schrödinger equation, on the other hand, focuses on the stationary states of a system, where the wave function doesn't change over time.

Imagine the time-dependent Schrödinger equation as a quantum "movie projector," playing out the evolving story of a quantum system frame by frame. Each frame represents the wave function at a particular moment, and the equation tells us how to transition from one frame to the next. The time-independent Schrödinger equation, in contrast, is like a quantum "photograph," capturing the unchanging essence of a stationary state.

Solutions and Implications

Solving the Schrödinger equation is no easy feat, but it's essential for understanding the behavior of quantum systems. Wave functions, the solutions to the equation, contain all the information about the system's properties and probabilities.

For example, solving the Schrödinger equation for a hydrogen atom reveals the electron's possible energy levels and orbitals. These solutions give us a glimpse into the intricate dance of subatomic particles and help explain the atomic spectra we observe in nature.

Picture the wave function as a quantum "blueprint," detailing the possible configurations and properties of a system. Just as an architect's blueprint guides the construction of a building, the wave function guides our understanding of the quantum world. By solving the Schrödinger equation, we can "read" these blueprints and uncover the secrets of atoms, molecules, and beyond.

Applications

The power of the Schrödinger equation extends far beyond theoretical physics. It has many practical applications in chemistry, materials science and electronics.

One notable example is the design of quantum dots, tiny semiconductor structures that exhibit quantum confinement effects. By solving the Schrödinger equation for these systems, researchers can predict and optimize their electronic and optical properties, paving the way for next-generation technologies like quantum computers and solar cells.

Imagine quantum dots as tiny "quantum paintbrushes," each capable of emitting light of a specific color. Scientists can create intricate nanoscale structures with tailored properties by carefully "painting" with these quantum dots. The Schrödinger equation acts as the "color palette," guiding the selection and arrangement of these quantum building blocks.

2.4: Operators and Eigenvalues - A Beginner's Guide

Quantum Operators Explained

In the quantum world, physical quantities like position, momentum, and energy are not just simple numbers; they are represented by mathematical objects called operators. These operators act on wave functions to extract information about the system's properties.

Imagine operators as "quantum probes" that interact with the wave function to reveal a system's hidden characteristics. Just as a doctor uses various instruments to examine a patient, quantum operators "examine" the wave function to uncover the system's physical attributes.

Let's consider the position operator, denoted as x, to understand how operators work. When applied to a wave function $\psi(x)$, the position operator returns the position coordinate x multiplied by the wave function:

$x\psi(x) = x\psi(x)$

This operation might seem simple, but it has profound implications. It tells us that the wave function $\psi(x)$ is an eigenfunction of the position operator, with the position coordinate x as its eigenvalue.

Eigenvalues and Physical Measurements

Eigenvalues and eigenfunctions are central concepts in quantum mechanics. An eigenfunction is a particular wave function that returns a multiple of itself when acted upon by an operator. The multiplying factor is called the eigenvalue.

In physical measurements, eigenvalues represent the possible outcomes of measuring a particular property. When we measure an observable, such as position or energy, we ask the corresponding operator to act on the wave function. The result is an eigenvalue, which tells us the observable value.

Picture eigenvalues as the "quantum price tags" attached to each eigenfunction. Just as a price tag tells you the cost of an item in a store, an eigenvalue tells you the value of an observable when the system is in the corresponding eigenfunction state.

Commutation and Uncertainty

One of the most intriguing aspects of quantum operators is their commutation behavior. Two operators are said to commute if the order in which they are applied doesn't affect the result. In other words, if A and B are commuting operators, then:

AB = BA

However, not all operators commute. Some of the most critical operators in quantum mechanics, such as position and momentum, do not commute. This non-commutation has a profound consequence: the Heisenberg uncertainty principle.

The uncertainty principle states that specific pairs of physical quantities, like position and momentum, cannot be simultaneously measured with arbitrary precision. The more precisely we know one quantity, the less precisely we can know the other.

Imagine a quantum "seesaw" with position on one end and momentum on the other. The uncertainty principle tells us that we can't balance both ends perfectly. If we push down on the position end, the momentum end rises, and vice versa. The more we try to pin down one quantity, the more the other one eludes us.

Using Operators in Quantum Mechanics

Operators are the workhorses of quantum mechanics. They allow us to calculate observables, predict the outcomes of measurements, and describe the evolution of quantum systems.

For example, the Hamiltonian operator denoted as $\hat{H}$ represents a system's total energy. We can find the system's energy levels and stationary states by applying the Hamiltonian operator to a wave function and solving the resulting Schrödinger equation.

Picture the Hamiltonian operator as a quantum "energy meter." Just as a regular energy meter measures the energy consumption of an electrical appliance, the Hamiltonian operator measures the energy of a quantum system. By "plugging in" different wave functions, we can determine the energy landscape of atoms, molecules, and beyond.

2.5: The Schrödinger Equation

Breaking Down the Schrödinger Equation

The Schrödinger equation lies at the center of quantum mechanics, a mathematical statement that encapsulates the behavior of quantum systems. This equation, developed by Austrian physicist Erwin Schrödinger in 1925, comes in two forms: the time-dependent Schrödinger equation and the time-independent Schrödinger equation.

The time-dependent Schrödinger equation describes how the wave function of a quantum system evolves. It takes into account the kinetic and potential energy of the system and relates them to the rate of change of the wave function. Mathematically, it is expressed as:

$$i\hbar \, \partial \psi(x,t) / \partial t = \hat{H} \psi(x,t)$$

where i is the imaginary unit, $\hbar$ (h-bar) is the reduced Planck's constant, $\psi(x,t)$ is the wave function, and $\hat{H}$ is the Hamiltonian operator, representing the system's total energy.

Imagine the time-dependent Schrödinger equation as a quantum "movie projector," playing out the evolving story of a quantum system frame by frame. Each frame represents the wave function at a particular moment, and the equation tells us how to transition from one frame to the next, guided by the system's energy and potential.

On the other hand, the time-independent Schrödinger equation focuses on the stationary states of a quantum system, where the wave function doesn't change over time. It is written as:

$$\hat{H} \psi(x) = E \psi(x)$$

Here, E represents the energy of the stationary state, and $\psi(x)$ is the time-independent wave function. This equation is an eigenvalue equation, where the wave function is an eigenfunction of the Hamiltonian operator, and the energy E is the corresponding eigenvalue.

Picture the time-independent Schrödinger equation as a quantum "still photograph," capturing the unchanging essence of a stationary state. It's like a snapshot of the system's energy landscape, revealing the possible energy levels and their associated wave functions.

Time-Dependent Schrödinger Equation

The time-dependent Schrödinger equation is the more general form, describing the temporal evolution of a quantum system. It allows us to predict how the wave function changes over time, considering the system's initial state and any external influences.

Solving the time-dependent Schrödinger equation can be complex, often requiring advanced mathematical techniques. However, the solutions provide a wealth of information about the system's behavior, such as the probability of finding the particle at different positions and times.

Imagine a quantum "dance floor," where the wave function represents the dancer's movements. The time-dependent Schrödinger equation acts as the "choreographer," guiding the dancer's steps and twirls as time progresses. By solving this equation, we can predict the dancer's position and momentum at any given moment, creating a beautiful and intricate quantum ballet.

Time-Independent Schrödinger Equation

The time-independent Schrödinger equation, on the other hand, is used to find the stationary states and energy levels of a quantum system. These stationary states are the backbone of quantum mechanics, as they represent the stable, unchanging configurations of the system.

By solving the time-independent Schrödinger equation, we can obtain the wave functions and energies of these stationary states. This information is crucial for understanding the structure of atoms, molecules, and solids and for predicting the outcomes of spectroscopic experiments.

Picture the time-independent Schrödinger equation as a quantum "energy map," charting the possible energy levels of a system. Each stationary state is like a landmark on this map, with its associated wave function describing the "shape" of the energy landscape at that point. By exploring this map, we can uncover the intricate energy structure of the quantum world.

Practical Solving Techniques

Solving the Schrödinger equation, whether in its time-dependent or time-independent form, is a crucial skill in quantum mechanics. While analytical solutions are possible for simple systems, such as the hydrogen atom or the particle in a box, more complex systems often require numerical methods.

One common technique is the variational method, which involves making an educated guess about the wave function and optimizing it to minimize energy. This method is beneficial for approximating a system's ground state energy and wave function.

Another approach is perturbation theory, which is used when the system of interest is similar to a solvable system but with a small additional term in the

Hamiltonian. The perturbation theory allows us to calculate the corrections to both the energy levels and wave functions.

Imagine the Schrödinger equation as a quantum "puzzle," with the wave function as the missing piece. The variational method is like making an educated guess about the shape of the missing piece based on the puzzle's overall structure. We then refine this guess, trying to find the best fit. Perturbation theory, on the other hand, is like starting with a completed puzzle and figuring out how a small change, like a slightly different color or shape, affects the overall picture.

2.6: Dirac Notation

Introduction to Dirac Notation

As we delve deeper into the mathematical formalism of quantum mechanics, we must familiarize ourselves with Dirac notation, also known as bra-ket notation. Developed by British physicist Paul Dirac, this elegant and compact notation has become the standard language for expressing quantum states and operators.

In Dirac notation, a quantum state is represented by a "ket" vector, denoted as $|\psi\rangle$. The "|" symbol is called a "ket," and the Greek letter ψ represents the specific state. For example, $|\psi_1\rangle$ and $|\psi_2\rangle$ represent two different quantum states.

Imagine a quantum "alphabet," where each letter represents a different state. The ket vectors are like the "words" formed from these letters, describing the various configurations of the quantum system. Just as words convey meaning in a language, ket vectors encapsulate the physical information about a quantum state.

The dual of a ket vector is called a "bra" vector, denoted as $\langle\psi|$. The "⟨" symbol is called a "bra," representing the complex conjugate of the corresponding ket vector. Bra vectors define inner products and calculate probabilities in quantum mechanics.

Bra and Ket Vectors

Bra and ket vectors are the building blocks of Dirac notation. They provide a concise way to represent the quantum states and their dual vectors, which are essential for performing calculations in quantum mechanics.

One of the key properties of bra and ket vectors is their orthonormality. If two states $|\psi_1\rangle$ and $|\psi_2\rangle$ are orthonormal, their inner product is defined as:

$$\langle\psi_1|\psi_2\rangle = \delta_{12}$$

where δ_{12} is the Kronecker delta equals 1 if the states are the same (i.e., 1 = 2) and 0 otherwise. This orthonormal condition is crucial for defining a basis set of states, which forms a complete description of the quantum system.

Picture bra and ket vectors as quantum "puzzle pieces." Each piece represents a different state, and when two pieces fit together perfectly, they form an orthonormal pair. The inner product is like the "click" you hear when two puzzle pieces snap into place, confirming their compatibility. A complete set of orthonormal states is like a finished puzzle, providing a comprehensive picture of the quantum system.

Inner and Outer Products

In Dirac notation, the inner product of two states $|\psi_1\rangle$ and $|\psi_2\rangle$ is written as $\langle\psi_1|\psi_2\rangle$. This inner product is a complex number that represents the overlap or projection of one state onto another. It is calculated by multiplying the bra vector $\langle\psi_1|$ with the ket vector $|\psi_2\rangle$.

The probability of measuring a quantum system in a particular state $|\psi_1\rangle$, given that it is initially in state $|\psi_2\rangle$, is equal to the absolute square of their inner product:

$P(\psi_1|\psi_2) = |\langle\psi_1|\psi_2\rangle|^2$

This relationship between inner products and probabilities is a fundamental quantum mechanics principle known as Born's rule.

Imagine the inner product as a quantum "compatibility test." It measures how well two states match or overlap with each other. The higher the absolute square of the inner product, the more compatible the states are, and the higher the probability of finding the system in one state given the other. It's like comparing two musical notes: if they are in harmony, their "inner product" is high, and they blend seamlessly.

The outer product, denoted as $|\psi_1\rangle\langle\psi_2|$, is another critical operation in Dirac notation. It creates a linear operator that maps one state to another. Outer products are used to construct projection operators, which project a state onto a specific subspace, and density matrices, which describe the statistical properties of a quantum system.

Advantages in Quantum Calculations

Dirac notation offers several advantages when performing quantum mechanical calculations. Its compact and intuitive form makes manipulating and simplifying expressions easier, especially when dealing with complex systems and transformations.

One key advantage of Dirac notation is its ability to express quantum states and operators independently of the choice of basis. This basis independence allows for greater flexibility in calculations and makes it easier to switch between different system representations.

Picture Dirac notation as a quantum "toolbox." Each tool (bras, kets, inner products, outer products) has a particular purpose and can be combined in various ways to solve quantum problems. Just as a skilled craftsman chooses the right tool for the job, a quantum physicist selects the appropriate Dirac notation elements to simplify calculations and uncover the system's properties. The versatility and adaptability of Dirac notation make it an indispensable tool in the quantum mechanic's arsenal.

Another advantage of Dirac notation is its ability to express the quantum state of a system as a linear combination of basis states. This superposition principle, which is fundamental to quantum mechanics, is easily represented using ket vectors:

$$|\psi\rangle = c_1|\psi_1\rangle + c_2|\psi_2\rangle + ... + c_n|\psi_n\rangle$$

where $|\psi\rangle$ is the quantum state, $|\psi_n\rangle$ are the basis states, and c_n are complex coefficients representing the amplitude and phase of each basis state's contribution.

Conclusion

In this chapter, we have explored the mathematical symphony of quantum mechanics, where various tools and concepts work together to create a beautiful framework for describing the quantum world.

Linear algebra, with its vectors and matrices, provides the backbone for expressing quantum states and operations. Complex numbers capture the wave-like nature of quantum entities, while differential equations, such as the Schrödinger equation, govern the dynamics of quantum systems. Operators and eigenvalues connect the abstract wave function to measurable physical quantities, and the Heisenberg uncertainty principle emerges from their commutation properties. Dirac notation ties it all together, offering a concise and intuitive way to perform quantum calculations.

Each mathematical concept plays a crucial role in the grand quantum mechanics puzzle, creating a harmonious and awe-inspiring whole. As we continue to explore the quantum world, these tools will guide us through its intricacies, helping us unravel the mysteries of subatomic particles and unlock the potential for groundbreaking technologies.

Beyond their practical applications, these mathematical concepts also offer a glimpse into the fundamental nature of reality, revealing a strange and beautiful world where particles behave like waves and probabilities reign supreme. With the knowledge acquired in this chapter, we are well-equipped to embark on this exhilarating journey of exploration into the captivating realm of quantum mechanics.

3

Beyond Classical Boundaries

Chapter 3: Beyond Classical Boundaries

This chapter will explore the fundamental principles of quantum mechanics and their mind-bending applications. From the enigmatic principle of superposition to the peculiar phenomenon of quantum tunneling, from the unbreakable codes of quantum cryptography to the perplexing quantum Zeno effect, we'll unravel the mysteries beyond the classical boundaries.

We'll also grapple with the philosophical implications of wave function collapse and explore the role of decoherence in bridging the quantum and classical worlds.

3.1: The Principle of Superposition and Its Applications in Quantum Computing

Understanding Superposition

Imagine a world where a coin can simultaneously be heads and tails, and a cat can be alive and dead until observed. Welcome to the bizarre realm of quantum superposition! In the quantum world, particles can exist in multiple states simultaneously, defying our classical intuition.

Think of a quantum particle as a spinner in a game of chance. In the classical world, the spinner can only point to one color at a time, either red or blue. But in the quantum world, the spinner can simultaneously point to red and blue, existing in a superposition of states. Only when we observe the spinner does it " collapse" into one definite color.

Superposition is like a quantum dance, where particles pirouette between different states until the music stops and they take their final pose. It's a choreography that challenges our understanding of reality and opens up a world of possibilities.

Quantum Bits (Qubits)

The idea of quantum bits, or qubits, is at the heart of quantum computing.

Unlike classical bits, which can only be 0 or 1, qubits can simultaneously exist in a superposition of 0 and 1. Quantum computers possess a unique property that enables them to perform specific calculations exponentially faster than traditional computers.

Imagine a classical bit as a light switch that can only be on or off. In contrast, a qubit is like a dimmer switch between fully on and fully off in any intermediate state. This analogy captures the essence of superposition, where a qubit can be in a continuum of states, not just binary ones.

Qubits are the elemental units of quantum computing, much like bits are to classical computing. But qubits are far more versatile and powerful, thanks to their ability to harness superposition and entanglement.

Quantum Computing

The principle of superposition is the bedrock of quantum computing, a revolutionary field that promises to solve problems intractable for classical computers. By leveraging the power of qubits and quantum algorithms, quantum computers can tackle challenges in cryptography, drug discovery, optimization, and more.

Think of a classical computer as a solo performer, methodically solving problems one step at a time. In contrast, a quantum computer is like a symphony orchestra, with each qubit playing its part in a grand, interconnected performance. The principle of superposition allows these qubits to explore multiple solutions simultaneously, leading to a quantum speedup.

Quantum computing is rapidly advancing with real-world applications; it's not a theoretical concept. Companies like Google, IBM, and Microsoft are investing heavily in quantum computing research, and the race is on to build the first practical, large-scale quantum computer.

Potential and Challenges

Quantum computing's potential is exciting and vast, but the path to realizing it is paved with challenges. One of the main hurdles is maintaining the delicate superposition of qubits, which can easily be disrupted by external noise and interference, leading to errors.

Just like trying to balance a pencil on its tip. The slightest disturbance can cause the pencil to fall, just like how the slightest interaction with the environment can cause a qubit to lose its superposition. This fragility of qubits is a significant challenge in building robust quantum computers.

Despite the challenges, researchers are making significant strides in developing quantum error correction techniques and improving the stability of qubits. The future of quantum computing is bright, and the potential benefits are too great to ignore.

3.2: Quantum Tunneling in Semiconductors and Technological Impacts

Basics of Quantum Tunneling

Imagine a ball rolling towards a hill. In the classical world, if the ball doesn't have enough energy to climb the hill, it will simply roll back. But in the quantum world, particles can do something strange: tunnel through barriers they shouldn't be able to cross according to classical physics. This phenomenon is known as quantum tunneling.

Imagine a quantum particle as a ghost that can pass through walls. In the quantum world, particles lack a definite position. Instead, they are described by a wave function that informs us about the likelihood of finding the particle in any given location. This wave function can expand beyond potential barriers, permitting particles to tunnel through them.

Tunneling in Semiconductors

Quantum tunneling is essential in semiconductor devices. Electrons can tunnel through thin insulating barriers, enabling current flow in devices like tunnel diodes and flash memory.

Imagine a semiconductor device as a maze with walls that electrons need to navigate. In classical electronics, the electrons would have to climb over these walls, requiring a certain amount of energy. But thanks to quantum tunneling, the electrons can sneak through the walls, allowing current to flow even when the classical conditions for conduction are not met.

Applications in Technology

Quantum tunneling's applications extend far beyond semiconductor devices. Tunneling is a critical mechanism in scanning tunneling microscopy (STM), a powerful technique for imaging surfaces at the atomic level. In STM, a sharp tip is brought close to a surface, and electrons tunnel between the tip and the surface, producing a detailed map of the atomic landscape.

Quantum tunneling also plays a role in the operation of superconducting quantum interference devices (SQUIDs), which measure fragile magnetic fields. SQUIDs rely on tunneling electron pairs across Josephson junctions, enabling them to detect magnetic fields with unparalleled sensitivity.

Future Technological Prospects

As our understanding of quantum tunneling deepens, the prospects for future technological applications are vast and exciting. Researchers are exploring tunneling in developing ultra-efficient solar cells, where electrons could tunnel through energy barriers to generate electricity more effectively.

Imagine a future where quantum tunneling is harnessed to create tiny, ultra-fast transistors that can power the next generation of computers. Or picture a world where tunneling-based sensors can detect single molecules, revolutionizing fields like medical diagnostics and environmental monitoring.

The study of quantum tunneling is also crucial for developing quantum computers. Tunneling is a key mechanism in the operation of superconducting qubits, one of the leading approaches to building scalable quantum processors. Researchers can create more robust and reliable qubits by understanding and controlling tunneling in these systems.

3.3: Quantum Entanglement in Cryptography

The Nature of Quantum Entanglement

Quantum entanglement is a phenomenon that defies our classical intuition. When two particles become entangled, they form a connection that transcends space and time. No matter how far apart these particles are, they remain linked in a way that classical physics cannot describe.

Imagine a pair of entangled particles as two lovers separated by a vast distance. Despite the physical separation, their hearts beat as one, and the other instantly feels any change in one lover's state. This mysterious connection, which Einstein famously called "spooky action at a distance," is the essence of quantum entanglement.

Entanglement has been experimentally verified countless times and is not a theoretical concept. Entanglement is a critical resource in quantum technologies, from quantum computing to quantum cryptography.

Principles of Quantum Cryptography

Quantum cryptography relies on the principles of quantum mechanics, especially entanglement, to secure communication. In classical cryptography, communication security depends on the computational difficulty of encryption cracking. However, in quantum cryptography, security is guaranteed by the laws of physics.

Imagine trying to eavesdrop on a conversation between two people speaking a language that changes every time you want to listen in. That's essentially how quantum cryptography works. By encoding information in the quantum states of entangled particles, any attempt to intercept the message will disturb the system, alerting the communicating parties to the presence of an eavesdropper.

Quantum cryptography has been implemented in real-world settings. For example, quantum-critical distribution networks have been established in several countries, enabling secure communication between banks, government agencies, and other organizations.

Quantum Key Distribution (QKD)

Quantum cryptography is based on the quantum key distribution protocol (QKD). This protocol involves two parties, usually called Alice and Bob, who create a shared encryption key using a series of entangled particles. The great thing about QKD is that if someone tries to intercept the key, the entanglement will be disturbed, making it detectable.

Imagine Alice and Bob as two spies who need to exchange secret messages. To ensure the security of their communication, they use a unique code book generated using entangled particles. If anyone tries to steal the code book, it will self-destruct, leaving behind a telltale sign of tampering.

QKD is not just limited to generating encryption keys; it can also be used for secure authentication and digital signatures, providing a theoretically unbreakable security level.

Implications for Security and Privacy

The emergence of quantum cryptography has significant implications for security and privacy in the digital age. As classical encryption methods become increasingly vulnerable to advances in computing power, quantum cryptography offers a future-proof solution immune to such threats.

Imagine a world where sensitive information, from financial transactions to medical records, is protected by the unbreakable laws of quantum mechanics. Quantum cryptography could revolutionize data and communication security, providing unprecedented privacy and protection.

However, implementing quantum cryptography also faces challenges, such as the need for dedicated infrastructure and the potential for side-channel attacks. As quantum cryptography research progresses, addressing these challenges is crucial for realizing its potential.

3.4: The Quantum Zeno Effect

Understanding the Quantum Zeno Effect

In the quantum world, observation has a profound effect on particle behavior. The quantum Zeno effect is a prime example, demonstrating how frequent measurements can "freeze" the evolution of a quantum system, preventing it from changing its state.

Imagine a watched pot that never boils. In the classical world, the act of observing the pot does not affect the boiling process. But in the quantum realm, if we keep checking on a particle's state, we can prevent it from evolving, just like the proverbial watched pot.

The quantum Zeno effect is named after the Greek philosopher Zeno, famous for his motion and change paradoxes. Just like Zeno's paradoxes challenge our intuition about the nature of reality, the quantum Zeno effect challenges our understanding of the relationship between observation and the quantum world.

The Role of Measurement in Quantum Mechanics

The quantum Zeno effect is a stark reminder of measurement's central role in quantum mechanics. In the quantum world, particles only have definite properties once observed. The act of measurement not only reveals the state of the particle but also shapes it.

Imagine a quantum particle as a chameleon that changes color depending on its environment. Every time we observe the chameleon, we force it to choose a color, influencing its state. The more frequently we observe the chameleon, the less likely it is to change color, just like how frequent measurements can suppress the evolution of a quantum system.

Experimental Evidence

The quantum Zeno effect has been demonstrated experimentally in various systems, ranging from trapped ions to superconducting qubits. These experiments showcase researchers' remarkable control over individual quantum systems and provide a glimpse into the fundamental nature of measurement in quantum mechanics.

One famous experiment demonstrating the quantum Zeno effect involved the suppression of atomic decay. Researchers could significantly extend an unstable atom's lifetime by repeatedly measuring its state, effectively "freezing" it in its initial state.

Philosophical and Practical Implications

The quantum Zeno effect has profound philosophical implications for our understanding of reality and the role of observation in shaping it. It challenges our classical notion of an objective reality that exists independently of our measurements and highlights the participatory nature of the quantum world.

From a practical perspective, the quantum Zeno effect has possible applications in quantum computing and quantum error correction. By leveraging the suppression of state evolution, researchers can develop techniques to preserve the delicate quantum states of qubits, extending their coherence times and reducing errors.

The quantum Zeno effect also has implications for our understanding of the arrow of time. In the quantum world, the direction of time's arrow is not as clear-cut as in the classical world. The suppression of state evolution by frequent measurements suggests that the flow of time in the quantum realm may be more malleable than we previously thought.

3.5: Wave Function Collapse

The Concept of Wave Function Collapse

At the center of quantum mechanics is the wave function, a mathematical object that encodes all the information about a quantum system. The Schrödinger equation governs the smooth and deterministic evolution of the particle's wave function, representing the probabilities of its various states. However, something strange happens when we make a measurement: the wave function seemingly "collapses" into a single definite state.

Imagine a quantum particle as a spinning top in multiple orientations simultaneously. The wave function describes the likelihood of finding the top in each orientation. When we measure the orientation of the top, it suddenly snaps into a single definite position, as if the act of observation forces it to choose.

The concept of wave function collapse is one of the most puzzling and controversial aspects of quantum mechanics. It implies that the act of observation has a special status in the quantum world, a notion that many physicists find unsatisfactory.

Debates and Interpretations

The interpretation of wave function collapse has been intensely debated among physicists and philosophers. Some, like the proponents of the Copenhagen interpretation, view collapse as a fundamental aspect of quantum mechanics. In contrast, others, like the advocates of the many-worlds interpretation, argue that collapse is an illusion and that all possible measurement outcomes occur in parallel universes.

Imagine a quantum detective trying to solve the mystery of wave function collapse. Each interpretation is like a different suspect with its own story. The Copenhagen interpretation points to the measurement process as the culprit. In contrast, the many-worlds interpretation suggests that the crime never happened and that all the suspects are guilty in their parallel realities.

Experimental Insights

Experiments probing the nature of wave function collapse have provided valuable insights into the quantum measurement process. For example, experiments with entangled particles have demonstrated that collapse occurs instantaneously and non-locally, defying our classical intuition about the speed of information transfer.

Recent experiments have also explored the possibility of observing the gradual collapse of the wave function in real time, shedding light on the dynamics of the measurement process. These experiments use weak measurements that only slightly disturb the quantum system, allowing researchers to track the evolution of the wave function during collapse.

Implications for Understanding Quantum Mechanics

The debate regarding wave function collapse is at the heart of our comprehension of quantum mechanics and the nature of reality. It challenges our classical notions of determinism, locality, and objectivity and forces us to confront the enigmatic role of the observer in the quantum world.

Imagine a quantum universe where our observations weave the very fabric of reality. In this universe, measurement actively participates in reality's creation, not just passively recording properties. Whether real or imaginary, the wave function collapse reminds us that the quantum world operates according to principles fundamentally different from our everyday experience.

The study of wave function collapse also has practical implications for developing quantum technologies. Understanding the nature of collapse is crucial for designing quantum measurement schemes and error correction protocols to preserve qubits' delicate quantum states.

3.6: Decoherence

Explaining Quantum Decoherence

Quantum decoherence is the phenomenon where a quantum system loses its coherence and begins to behave classically. It occurs when a quantum system interacts with its environment, becoming entangled with it. This interaction causes the distinct quantum states of the system to become mixed, leading to a loss of coherence and the emergence of classical behavior.

Imagine a pristine quantum world as a calm pond where the ripples of probability maintain their coherence and interference. Decoherence is like a gust of wind that disturbs the pond, causing the ripples to become chaotic and disorderly. As the wind blows, the quantum pond starts to resemble a classical pond, with no trace of its former quantum coherence.

Decoherence is not a unitary process like the evolution of the wave function under the Schrödinger equation. Instead, it is an irreversible process that increases the entropy of the quantum system, leading to a loss of information and the emergence of classical probabilities.

Decoherence and Measurement

Decoherence links the quantum world and the classical world of measurements. When we measure a quantum system, we effectively introduce decoherence by coupling the system to a macroscopic measuring device. This interaction causes the quantum states of the system to become entangled with the states of the measuring device, leading to a loss of coherence and the appearance of definite measurement outcomes.

Imagine a quantum magician performing a trick with a coherent quantum coin. As soon as the magician interacts with the coin, either by observing it directly or by using a measuring device, the coin loses its coherence, and it behaves like a classical coin with a definite outcome.

The role of decoherence in measurement has led some physicists to propose that decoherence can explain the appearance of wave function collapse without the need for a separate collapse postulate. In this view, the apparent collapse of the wave function is a consequence of the rapid decoherence induced by the measuring device.

Experimental Observations

Compelling evidence exists for decoherence and its role in quantum-classical transition. For example, experiments with superconducting qubits have shown that decoherence can be suppressed by carefully engineering the qubit's environment, extending its coherence time and enabling longer quantum computations.

Other experiments have directly observed the process of decoherence in real-time, using techniques such as quantum state tomography. These experiments have demonstrated the gradual transition from quantum coherence to classical behavior when a quantum system interacts with its environment.

The Role of Decoherence in Quantum Computing

Decoherence is one of the biggest challenges facing the development of large-scale quantum computers. Quantum computing relies on maintaining coherent quantum states and performing quantum operations without the disruptive influence of decoherence. However, as quantum systems become larger and more complex, they become increasingly susceptible to decoherence, limiting the depth and accuracy of quantum computations.

Imagine a quantum orchestra playing a symphony, where each qubit is a musician playing in perfect harmony. Decoherence is like a noisy audience that disrupts the symphony, causing the musicians to lose their synchronization and the music to become dissonant. The challenge for quantum computing is to isolate the orchestra from the noisy audience and maintain its coherence for as long as possible.

Various strategies are being developed to combat decoherence in quantum computing. These include quantum error correction, topological quantum computing, and decoherence-free subspaces. These approaches aim to create quantum systems inherently resilient to decoherence, enabling reliable quantum computations even in environmental noise.

Conclusion

In this chapter, we have explored the fundamental principles and applications of quantum mechanics, revealing how the quantum world challenges our classical intuition and opens up new possibilities for technology and our understanding of reality.

From superposition and quantum tunneling to quantum cryptography and the quantum Zeno effect, the quantum realm is full of surprises and insights that defy our everyday experience. These concepts challenge our notions of reality, measurement, and the nature of the universe, inviting us to question our assumptions and embrace the strange and beautiful world of quantum mechanics.

We have also seen how quantum principles find practical applications in developing quantum computers, semiconductors, secure communication, and ultra-sensitive measurements. These applications remind us that the study of quantum mechanics is not just an abstract intellectual pursuit but a field with the potential to revolutionize technology and transform our world.

As we continue our exploration of quantum mechanics, it is essential to remember that there are still many open questions and unresolved puzzles, from the nature of quantum gravity to the origin of quantum randomness. Unravelling the universe's deepest secrets will require the combined efforts of scientists and philosophers alike.

Ultimately, studying quantum mechanics is a testament to the power of human curiosity and the enduring quest for knowledge. By embracing the strange and the counterintuitive, we open ourselves up to a world of endless possibilities and profound insights.

4

Quantum Mechanics at Work

Chapter 4: Quantum Mechanics at Work

In this chapter, we'll explore the practical aspects of quantum mechanics, from solving the enigmatic Schrödinger equation to understanding its role in chemistry, computing, and everyday life. We'll even explore some hands-on quantum experiments you can try at home, bringing the abstract concepts to life.

We'll also glimpse the future of quantum technologies, exploring the potential breakthroughs and challenges that await us as we unravel the quantum world's secrets.

4.1: Solving the Schrödinger Equation

Introduction to the Schrödinger Equation

The Schrödinger equation lies at the heart of quantum mechanics, a mathematical expression that encapsulates the behavior of quantum systems. This equation, named after the Austrian physicist Erwin Schrödinger, is the quantum equivalent of Newton's laws of motion in classical mechanics. It describes how the wave function, a mathematical object that contains all the information about a quantum system, evolves.

The Schrödinger equation is like a quantum recipe book. Just as a recipe tells you how to combine ingredients to create a dish, the Schrödinger equation tells you how to integrate the various components of a quantum system (such as potential energy and kinetic energy) to determine its wave function and, ultimately, its properties.

Solving Simple Systems

To understand how the Schrödinger equation works, let's start with a simple example: the hydrogen atom. This system consists of a single electron orbiting a proton, held together by the electromagnetic force. By solving the Schrödinger equation for the hydrogen atom, we can determine the allowed energy levels of the electron and the corresponding wave functions.

Solving the Schrödinger equation involves several mathematical techniques, such as separating variables and boundary conditions. These techniques allow us to break down the complex equation into more straightforward, more manageable parts.

Solving the Schrödinger equation is like solving a complex puzzle. You start by breaking it down into smaller, more manageable pieces (like separating the variables) and then work through each piece, applying the rules of quantum mechanics (like boundary conditions) until you arrive at the final solution.

Techniques and Approximations

As we move beyond simple systems like the hydrogen atom, solving the Schrödinger equation becomes increasingly challenging. In many cases, it is impossible to find exact solutions and we must rely on approximation methods.

One such method is perturbation theory, which allows us to find approximate solutions to the Schrödinger equation when the system is subject to a minor disturbance, or "perturbation." Another powerful technique is the variational method, which provides an upper bound on a system's ground state energy by minimizing the energy's expectation value concerning a trial wave function.

These approximation methods are like quantum shortcuts. They allow us to navigate the complex landscape of quantum systems without getting bogged down in the mathematical details. By making strategic simplifications and assumptions, we can arrive at solutions that are accurate enough for practical purposes, even if they are not exact.

Interpreting Solutions

Once we have solved the Schrödinger equation for a given system, the next step is to interpret the solutions. The wave functions that emerge from the Schrödinger equation contain a wealth of information about the quantum system, from the probability distribution of the particles to the allowed energy levels.

For example, in the case of the hydrogen atom, the wave functions correspond to the different atomic orbitals, such as the s, p, and d orbitals. These orbitals represent the probability distribution of the electron around the nucleus, and their shapes and energies determine the atom's chemical properties.

Interpreting the solutions of the Schrödinger equation is like decoding a quantum message. The wave functions are the language in which the quantum world communicates with us, and by learning to read this language, we can unlock the secrets of atoms, molecules, and beyond.

4.2: Quantum Mechanics in Chemistry

Quantum Chemistry Basics

Quantum mechanics is not just a physics theory but also the foundation of modern chemistry. The behavior of electrons, which determine the chemical properties of atoms and molecules, is governed by the laws of quantum mechanics. This realization led to the development of quantum chemistry, a field that applies quantum mechanics to study chemical systems.

Quantum chemistry is like a quantum microscope. It allows us to zoom in on the intricate dance of electrons that underlies chemical bonding and reactivity. By understanding the quantum mechanics of electrons, we can predict the properties of molecules and design new materials with tailored characteristics.

The Quantum Nature of Chemical Bonds

At the heart of quantum chemistry lies the concept of chemical bonding. In the quantum world, chemical bonds arise from the sharing or transferring of electrons between atoms. The wave functions of the electrons, which are solutions to the Schrödinger equation, determine the strength and properties of these bonds.

For example, in a covalent bond, two atoms share a pair of electrons, whose wave functions overlap to form a bonding orbital. The energy of this bonding orbital is lower than the energy of the individual atomic orbitals, which hold the atoms together. In an ionic bond, an electron is completely transferred from one atom to another, creating a pair of charged ions that attract each other through electrostatic forces.

Chemical bonding is like a quantum dance. With their wave-like properties, the electrons engage in a delicate choreography, swirling around the atomic nuclei and forming intricate patterns that give rise to the vast diversity of molecules and materials we observe in nature.

Applications in Molecular Chemistry

Quantum chemistry has numerous applications in the study of molecules and their properties. By solving the Schrödinger equation for molecular systems, we can predict molecules' geometry, energy and reactivity.

One of the most important applications of quantum chemistry is spectroscopy. We can learn about their electronic structure and bonding by studying how molecules absorb and emit light. Different types of spectroscopy, such as infrared, ultraviolet-visible, and nuclear magnetic resonance, probe various aspects of molecular structure and provide valuable information for chemical analysis.

Quantum chemistry is like a molecular detective agency. By analyzing the quantum fingerprints of molecules, such as their spectroscopic signatures, we can unravel the mysteries of their structure and behavior. This information is crucial for designing new drugs, developing advanced materials, and understanding the complex chemistry of life.

Computational Quantum Chemistry

As the complexity of chemical systems increases, solving the Schrödinger equation by hand becomes impractical. This is where computational quantum chemistry comes in. By harnessing the power of computers, we can perform quantum chemical calculations on large molecules and complex materials.

Computational quantum chemistry involves various methods, from ab initio techniques that solve the Schrödinger equation directly to semi-empirical methods that use experimental data to simplify the calculations. These methods allow us to predict the properties of molecules with increasing accuracy and efficiency.

Creating virtual molecules in a computer allows us to explore their properties and reactions without setting foot in a lab. This virtual experimentation will enable us to design new materials, optimize chemical processes, and even discover new drugs, all from the comfort of our computers.

4.3: Quantum Computing

Basics of Quantum Computing

Quantum computing operates based on the principles of quantum mechanics. Unlike classical computers, which rely on bits with only two values - 0 or 1- quantum computers use qubits, which can exist simultaneously in a superposition of 0 and 1.

Quantum computing is like parallel processing on steroids. By exploiting the principle of superposition, quantum computers can perform many calculations simultaneously, exploring multiple solutions in parallel. This quantum parallelism is what gives quantum computers their incredible speed and power.

Understanding Qubits

Quantum computers are built using physical systems such as atoms, ions, or superconducting circuits called qubits, which are manipulated using quantum operations. A qubit's state is described by a wave function, which encodes the probability amplitudes of being in the $|0\rangle$ or $|1\rangle$ state.

One of the key properties of qubits is entanglement. When two or more qubits are entangled, their quantum states correlate in a way that classical physics cannot describe. Entanglement allows quantum computers to perform tasks, such as teleportation and superdense coding, that are impossible with classical computers.

Qubits can perform incredible computation feats by leveraging superposition and entanglement principles. By carefully choreographing the interactions between qubits, we can create quantum algorithms that solve problems faster than any classical computer.

Quantum Algorithms

Quantum algorithms use qubits' unique properties, such as entanglement and superposition, to perform computations more efficiently than classical algorithms on quantum computers.

Shor's algorithm is one of the most renowned quantum algorithms, which can factorize large numbers exponentially faster than the best-known classical algorithm. This has significant implications for cryptography, as many modern encryption methods rely on the difficulty of factoring large numbers.

Another critical quantum algorithm is Grover's algorithm, which can search an unsorted database quadratically faster than classical algorithms. This speedup can be applied to various optimization problems, from drug discovery to machine learning.

Quantum algorithms allow us to perform computational feats that seem impossible from a classical perspective. By carefully exploiting the quantum properties of qubits, these algorithms can solve problems that would take centuries or even millennia on classical computers.

Challenges and Future Prospects

While quantum computing holds immense promise, significant challenges remain before we can build practical, large-scale quantum computers. One main challenge is maintaining qubits' coherence, which can easily be disrupted by environmental noise and errors.

Another challenge is scaling up quantum computers to include more qubits. The complexity of controlling and manipulating qubits increases with the number of qubits. Developing new technologies, such as quantum error correction and fault-tolerant quantum computing, will be crucial for realizing the full potential of quantum computers.

Despite the challenges, the prospects of quantum computing are incredibly exciting. From quantum cryptography and quantum simulation to quantum machine learning and quantum optimization, quantum computing's applications span a wide range of fields and industries.

The journey towards practical quantum computing is like a quantum odyssey. We are embarking on an epic adventure, full of challenges and opportunities, that will take us to new frontiers of computation and discovery. As we navigate this uncharted territory, we must be creative, persistent and bold, constantly pushing the boundaries of what is possible with quantum technology.

4.4: Quantum Mechanics in Everyday

Quantum Mechanics Around Us

Quantum mechanics may seem abstract and esoteric, far removed from everyday experience. However, the truth is that quantum mechanics plays a vital role in many of the technologies we use daily, from the smartphones in our pockets to the GPS systems in our cars.

Quantum mechanics is like the silent hero of modern technology. It works behind the scenes, powering many devices and systems we take for granted. Without quantum mechanics, our world would be very different, devoid of many conveniences and capabilities we have come to rely on.

Quantum Effects in Technology

One of the most striking examples of quantum mechanics in action is the Global Positioning System (GPS). GPS relies on satellites that orbit the planet, constantly broadcasting their position and time. A GPS receiver can accurately determine its location by comparing the signals from different satellites.

However, to achieve this accuracy, GPS must consider the effects of both special and general relativity. The clocks on the satellites, which are moving at high speeds and experiencing weaker gravitational fields than clocks on Earth, tick at a rate different from Earth-based clocks. If these relativistic effects were not accounted for, GPS would quickly become useless, with errors accumulating about 10 kilometers daily!

The role of quantum mechanics in GPS is like a cosmic ballet. With their precise atomic clocks, the satellites dance around the Earth in a delicate choreography, their movements and timings dictated by the laws of relativity. By carefully accounting for these relativistic effects, we can ensure that GPS remains a reliable and indispensable tool for navigation and positioning.

The Future of Quantum Technologies

As our understanding of quantum mechanics deepens, so does the potential for new and transformative technologies. One of the most exciting areas of development is quantum sensing, which exploits the exquisite sensitivity of quantum systems to probe the world around us with unprecedented precision.

Quantum sensors can potentially revolutionize fields as diverse as medical imaging, geological exploration, and fundamental physics. By harnessing the power of quantum entanglement and superposition, these sensors can detect tiny changes in magnetic fields, gravitational fields, and other physical quantities, opening up new frontiers of measurement and discovery.

Another promising area is quantum cryptography, which uses the principles of quantum mechanics to create unbreakable encryption methods. By encoding information in the quantum states of photons, quantum cryptography can ensure communications security against even the most sophisticated eavesdropping attempts.

As we push the boundaries of what is possible with quantum mechanics, we are opening up new vistas of opportunity and discovery. From quantum computers that can solve problems beyond the reach of classical machines to quantum sensors that can probe the universe's most profound mysteries, the potential of quantum technologies is limitless.

Bridging the Gap Between Quantum and Classical

As we explore the role of quantum mechanics in our everyday lives and contemplate the future of quantum technologies, we must remember that the quantum world is not separate from the classical world we inhabit. Rather, the two are inextricably linked, with quantum mechanics providing the foundation upon which our classical reality emerges.

By understanding the quantum underpinnings of the technologies we use daily, we gain a deeper understanding of the incredible beauty and complexity of the world around us. We begin to see that the seemingly mundane devices and systems that populate our lives are the product of a rich tapestry of quantum phenomena woven together by the laws of nature.

Narrowing the gap between the quantum and classical worlds requires confronting quantum mechanics' strange and counterintuitive aspects and finding ways to reconcile them with our everyday experience. By doing so, we gain a deeper understanding of our universe and open up new possibilities for harnessing the power of quantum mechanics to transform our world.

4.5: Practical Quantum Experiments You Can Try at Home

DIY Quantum Experiments

Learning about quantum mechanics doesn't have to be confined to textbooks and lectures. With creativity and a few simple materials, you can experience the wonders of the quantum world firsthand in your own home.

Doing quantum experiments at home is like being a quantum explorer. You get to venture into the strange and fascinating world of quantum mechanics, armed with nothing but your curiosity and a few household items. By performing these experiments, you can gain a tangible understanding of the quantum phenomena that underlie our universe.

One of the most accessible quantum experiments is the double-slit experiment, which demonstrates the wave-particle duality of light. All you need is a laser pointer, a piece of cardboard with two slits cut into it, and a white wall or screen. By shining the laser through the slits and observing the interference pattern that forms on the screen, you can see firsthand how light behaves like a wave, even though it is composed of individual photons.

Another fun experiment is the quantum eraser experiment, which explores the role of measurement in quantum mechanics. By using polarizing filters and a source of entangled photons, you can demonstrate how the act of measuring one photon can instantaneously affect the behavior of its entangled partner, even if a significant distance separates the two photons.

The quantum eraser experiment shows how the act of measurement can seemingly erase information about a quantum system and how this erasure can have instantaneous effects on other parts of the system. By performing this experiment, you can better appreciate quantum entanglement's strange and counterintuitive nature.

Understanding Through Doing

The value of doing quantum experiments at home goes beyond mere entertainment. By actively engaging with the phenomena of quantum mechanics, you can develop a deeper and more intuitive understanding of the concepts that underlie this fascinating field.

Hands-on experimentation allows you to confront the counterintuitive aspects of quantum mechanics and grapple with the implications of these phenomena for our understanding of reality. By seeing the results of these experiments with your own eyes, you can begin to internalize the wonderful and strange world of quantum mechanics in a way that reading about it alone cannot provide.

Doing quantum experiments is like learning a new language. At first, the concepts and terminology may seem foreign and need clarification. But by immersing yourself in the language through active practice and experimentation, you can gradually develop fluency and mastery. In the same way, by doing quantum experiments, you can become a native speaker of the language of quantum mechanics.

Connecting Theory to Practice

Of course, quantum experiments are not just about playing with lasers and polarizers. To truly benefit from them, it is vital to connect them back to the underlying theoretical principles of quantum mechanics.

Each experiment should be accompanied by a clear explanation of the quantum concepts it demonstrates and how these concepts fit into the broader framework of quantum theory. By making these connections explicit, you can use your experimental observations to reinforce your conceptual understanding and gain new insights into the subtle and often surprising implications of quantum mechanics.

Inspiring Further Exploration

The most significant value of doing quantum experiments at home is that it can inspire a lifelong passion for learning and discovery. By experiencing the wonders of quantum mechanics firsthand, you may find yourself drawn to explore this fascinating field in greater depth, and to pursue further studies in physics, mathematics, or related disciplines.

Moreover, the skills and mindset developed through hands-on experimentation are valuable beyond quantum mechanics. You can approach any problem or challenge with the tools and confidence needed to succeed by cultivating a spirit of curiosity, creativity and critical thinking.

4.6: The Future of Quantum Technologies

Quantum Technology Roadmap

As we look to the future of quantum technologies, it is clear that we are on the cusp of a new era of innovation and discovery. From quantum computing and communication to quantum sensing and metrology, the potential applications of quantum mechanics are vast and far-reaching.

The quantum technology roadmap is like a map of a new and uncharted territory. It lays out the key milestones and challenges ahead and provides a guide for navigating quantum technologies' complex and rapidly evolving landscape. By following this roadmap, we can chart a course towards a future in which quantum mechanics plays a pivotal role in our lives.

In the coming years and decades, we can expect significant advances in quantum computing hardware and software, with the development of larger and more powerful quantum processors and new algorithms and applications that harness these devices' unique capabilities. We may also see the emergence of quantum communication networks that enable secure and efficient communication over long distances and quantum sensors that can detect and measure physical quantities with unprecedented precision and sensitivity.

Impact on Society

As quantum technologies mature and become more widespread, they are expected to significantly impact society, revolutionizing industries, economies and daily lives.

In healthcare, for example, quantum sensors could enable earlier and more accurate diagnosis of diseases, while quantum computing could propel the discovery of new drugs and therapies. In finance, quantum algorithms could optimize portfolio management and risk assessment, while quantum cryptography could provide unbreakable security for financial transactions. In energy and sustainability, quantum technologies could help us develop more efficient solar cells, batteries, and other clean energy solutions.

The impact of quantum technologies on society is like a quantum butterfly effect. Just as small changes in initial conditions can lead to significant and unpredictable effects in chaotic systems, the development of quantum technologies could have far-reaching and transformative consequences for our world. By anticipating and preparing for these changes, we can work to ensure that the benefits of quantum technologies are widely shared and that their risks and challenges are effectively managed.

At the same time, the development of quantum technologies also raises important ethical and societal questions that we will need to grapple with as a global community. Privacy, security and equity issues will become increasingly crucial as quantum technologies become more prevalent. We will need to develop new policies and frameworks and ensure that these technologies are used responsibly and for the benefit of all.

Preparing for a Quantum Future

For those excited by the promise and potential of quantum technologies, there are many ways to get involved and prepare for the quantum future. Whether you are a student looking to pursue a career in quantum science and engineering or simply someone curious about this fascinating field, there are many opportunities to learn, explore, and contribute.

Embarking on a quantum journey requires curiosity, adaptability, and a commitment to pushing boundaries. By embracing the challenges and opportunities of the quantum world, we can all help mould the future of this exciting and rapidly evolving field.

One way to get started is to pursue education and training in quantum physics, mathematics, computer science, or related fields. Many research institutions and universities now offer courses and programs in quantum science and technology, and there are also many online resources and communities where you can learn and connect with others who share your interests.

Another way to get involved is to engage in research and development in quantum technologies through academic or industry partnerships. By working on cutting-edge projects and collaborating with experts from diverse fields, you can contribute to advancing quantum science and help bring new technologies and applications to life.

The Interdisciplinary Nature of Quantum Research

As we embark on this quantum journey, we must recognize that developing quantum technologies is inherently interdisciplinary, requiring collaboration and cooperation across many fields and sectors.

From computer science and physics to engineering and mathematics, from industry and government to academia and civil society, the success of quantum technologies will depend on our ability to work together and leverage the diverse skills, perspectives and resources of all stakeholders.

Just as quantum particles can become interconnected across vast distances, the insights and innovations that will drive the quantum future will emerge from the complex web of interactions and collaborations that span multiple fields and domains. By embracing this entanglement and working together across disciplinary boundaries, we can unlock the full potential of quantum technologies and create a brighter, more sustainable future for all.

Conclusion

In this chapter, we have seen how the principles of quantum mechanics are already shaping the technologies we use daily, from GPS to computer chips. However, this is just the beginning. As we continue to explore and harness the power of quantum mechanics, we stand on the threshold of a new era of innovation and discovery that promises to transform our world in ways we can only begin to imagine.

There has never been a more exciting time to be involved in quantum science and technology. By learning, experimenting, collaborating, and pushing the boundaries of what is possible, we can all play a role in shaping the quantum future and unlocking the incredible potential of this fascinating and rapidly evolving field.

Let us embrace the challenges and opportunities of the quantum world and work together to build a brighter, more sustainable, and more deeply connected future. The quantum future is waiting for us—let us rise to meet it with curiosity, creativity and courage.

5

Bridging Quantum Mechanics and Relativity

Chapter 5: Bridging Quantum Mechanics and Relativity

Now, it's time to embark on a journey that bridges the realms of quantum mechanics and relativity, two of the most profound and transformative theories in modern physics.

In this chapter, we'll explore the intersection of quantum mechanics and general relativity. We'll go through the history of their reconciliation, from the early equations to the advanced concepts of quantum field theory. We'll also discuss the quest for a theory of quantum gravity that unifies these two frameworks.

5.1: Special Relativity and Quantum Mechanics

Merging Frameworks

At first glance, quantum mechanics and special relativity may seem like two distinct and incompatible theories, each with its domain of applicability. Quantum mechanics, with its probabilistic description of the subatomic world, is at odds with the deterministic and geometric nature of special relativity, which deals with the behavior of space and time at high velocities.

However, upon closer inspection, we see that these two frameworks share a deep connection in redefining our notions of space, time, and causality. Both theories challenge our classical intuition and force us to rethink the nature of reality at its most fundamental level.

The merging of quantum mechanics and special relativity is like the convergence of two rivers, each with unique characteristics and flow patterns. As they come together, they create a new, more powerful stream that combines both properties and opens up possibilities for exploration and discovery.

Simultaneity and Time Dilation

One of the fundamental principles of special relativity is that the simultaneity of events can be relative. This means that what may seem simultaneous to one observer may not be simultaneous to another moving at a different velocity. This concept has profound implications for our understanding of time and causality and significant consequences for quantum mechanics.

In the quantum world, the notion of simultaneity becomes even more complex, as the probabilistic nature of quantum measurements means that the timing of events can be inherently uncertain. Moreover, the time dilation predicted by special relativity can affect the behavior of quantum systems, leading to novel phenomena such as the relativistic quantum Hall effect.

The interplay between simultaneity and time dilation in quantum mechanics is like a dance between two partners, each moving to their rhythm but ultimately influencing and adapting to each other's movements. The result is a beautiful and intricate choreography that pushes the boundaries of our understanding of space and time.

Relativistic Quantum Particles

The behavior of particles at high velocities, where relativistic effects become significant, is a crucial point of intersection between quantum mechanics and special relativity. In this regime, the non-relativistic Schrödinger equation can no longer describe particles, and we must turn to relativistic quantum equations such as the Klein-Gordon and Dirac equations.

These equations reveal a rich tapestry of phenomena, from the existence of antimatter to the concept of spin, which arises from the merger of quantum mechanics and special relativity. They also provide a foundation for the development of quantum field theory, which describes particles as excitations of underlying quantum fields and provides a unified framework for understanding the fundamental forces of nature.

Relativistic quantum particles are like cosmic dancers, moving at breathtaking speeds and performing incredible feats of acrobatics that defy our classical intuition. They embody the beauty and complexity of the universe at its most fundamental level, and they invite us to explore the frontiers of physics with a sense of wonder and curiosity.

Challenges and Solutions

Although relativistic quantum equations have successfully described the behavior of particles at high velocities, unifying quantum mechanics and special relativity remains challenging. One of the main difficulties is the apparent incompatibility between the instantaneous collapse of the wave function in quantum measurements and the finite speed of light in special relativity.

Another challenge is the problem of divergences, which arise when attempting to calculate specific quantities in relativistic quantum theories and lead to infinities that have no physical meaning. These divergences have been addressed through the development of renormalization techniques, which have proven incredibly successful in making accurate predictions in quantum field theory.

The challenges and solutions in merging quantum mechanics and special relativity are like a complex puzzle. Each piece represents a different theory aspect and requires a unique approach to fit into the larger picture. Solving this puzzle requires creativity, perseverance and a willingness to think outside the box.

5.2: The Klein-Gordon Equation

Birth of the Klein-Gordon Equation

The Klein-Gordon equation, developed in the 1920s by physicists Oskar Klein and Walter Gordon, was one of the first attempts to create a relativistic version of the Schrödinger equation. The goal was to find a wave equation consistent with quantum mechanics and special relativity.

The Klein-Gordon equation is a second-order partial differential equation that describes the behavior of a scalar field in relativistic space-time. It is derived by combining the energy-momentum relation from special relativity with the quantum mechanical operators for energy and momentum.

Interpreting the Equation

The Klein Gordon equation has several essential features distinguishing it from the non-relativistic Schrödinger equation. One of the most notable is the presence of negative energy solutions, which were initially interpreted as problematic and unphysical.

However, later developments in quantum field theory showed that these negative energy solutions could be reinterpreted as positive energy antiparticles, leading to the concept of antimatter. This realization was a major breakthrough in our understanding of the nature of matter and the universe.

Another key feature of the Klein-Gordon equation is its invariance under Lorentz transformations, which means that it is consistent with the principles of special relativity. This invariance ensures that the equation describes the same physical phenomena in all inertial reference frames.

Interpreting the Klein-Gordon equation is like deciphering a cosmic code, unlocking the universe's secrets one symbol at a time. Each term in the equation represents a different aspect of the physical world, from the energy and momentum of particles to the curvature of space-time itself. By learning to read and understand this code, we gain a deeper appreciation for the elegant mathematical language in which the laws of nature are written.

Applications and Limitations

Despite its groundbreaking nature, the Klein-Gordon equation has some limitations in describing certain types of particles. In particular, it fails to account for the intrinsic spin of particles like electrons, which is a fundamental quantum property with no classical analog.

Moreover, the Klein-Gordon equation predicts the existence of negative probability densities, which are difficult to interpret physically. These limitations led to the development of more advanced relativistic quantum equations, such as the Dirac equation, which provides a more complete description of relativistic quantum particles.

Nevertheless, the Klein-Gordon equation remains essential in modern physics, particularly in studying scalar fields and the Higgs boson. It also functions as a valuable pedagogical tool for introducing students to the concepts of relativistic quantum mechanics.

See the applications and limitations of the Klein-Gordon equation as the two sides of a coin, each revealing a different aspect of the theory's strengths and weaknesses. On one side, we see the equation's power to predict new phenomena and shed light on the nature of matter and energy. Conversely, we see its limitations and the need for further development and refinement. By studying both sides of the coin, we gain a more complete understanding of the role of the Klein-Gordon equation in the larger tapestry of physics.

Beyond the Klein-Gordon

The Klein-Gordon equation was a critical stepping stone in developing relativistic quantum mechanics, paving the way for more advanced theories such as the Dirac equation and quantum field theory. These theories build upon the foundation laid by the Klein-Gordon equation, extending its scope and power to describe a wider array of physical phenomena.

The Dirac equation, in particular, represents a breakthrough in our understanding of relativistic quantum particles. It provides a consistent description of particles with spin and predicts the existence of antimatter. It also serves as the basis for quantum electrodynamics, one of the most precisely tested theories in physics.

5.3: The Dirac Equation

Dirac's Revolutionary Idea

In the late 1920s, British physicist Paul Dirac was dissatisfied with the limitations of the Klein-Gordon equation and set out to find a more complete relativistic quantum equation. His goal was to develop a theory that would consistently describe the behavior of electrons, which are spin-1/2 particles, in a way that was compatible with both quantum mechanics and special relativity.

Dirac developed a new approach to describe the behavior of a quantum particle with spin by taking the square root of the relativistic energy-momentum relation. This led him to introduce a new mathematical object called a spinor. The final result is the Dirac equation, which comprises four coupled first-order partial differential equations. It is used to explain the behavior of a relativistic quantum particle with spin.

Dirac's revolutionary idea illuminated the path forward and revealed new horizons of possibility. It was a moment of pure creative genius, born of a deep understanding of the fundamental principles of physics and a willingness to think outside the box. The impact of Dirac's work on the development of modern physics cannot be overstated, and it continues to inspire and guide us to this day.

Implications of the Dirac Equation

The Dirac equation had several profound implications for understanding the nature of matter and the universe. Perhaps the most striking of these was the prediction of antimatter, which emerged naturally from the equation's negative energy solutions.

According to the Dirac equation, for every type of particle with positive energy, a corresponding antiparticle with negative energy must also exist. This means that for every electron, there must be a positron for every proton, an antiproton, and so on. The existence of antimatter was later confirmed experimentally, providing a stunning validation of Dirac's theory.

The implications of the Dirac equation are like ripples in a pond, spreading out from the initial point of impact and touching every corner of the physical world. They challenge our assumptions about the nature of reality and invite us to explore new realms of possibility, from the behavior of subatomic particles to the structure of the universe. By embracing the implications of the Dirac equation, we open ourselves up to a deeper understanding of the fundamental laws that govern the cosmos.

Understanding Spin

Another key feature of the Dirac equation is its description of the intrinsic spin of particles. In non-relativistic quantum mechanics, spin is introduced as an ad hoc assumption, without any deeper explanation of its origin or nature. The Dirac equation, on the other hand, naturally incorporates spin as a consequence of the theory's mathematical structure.

In the Dirac equation, spin emerges as a type of internal angular momentum inherent to the particle itself rather than the result of any external motion or rotation. This understanding of spin as a fundamental property of particles has become a cornerstone of modern particle physics.

The Dirac equation reminds us that the world is often more complex and subtle than we initially assume and that proper understanding requires a willingness to question our assumptions and explore new perspectives.

Impact on Particle Physics

The impact of the Dirac equation on the development of particle physics cannot be overstated. It provided a consistent framework for describing the behavior of relativistic quantum particles and laid the foundation for many of the key advances in the field over the past century.

The Dirac equation finds one of its most crucial applications in the development of quantum electrodynamics (QED), which is essentially the quantum theory of the electromagnetic interaction. QED is a highly precise theory in physics and its predictions have been tested to an accuracy of better than one part in a billion, making it one of the most accurately tested theories in physics.

The Dirac equation also played a crucial role in evolving the Standard Model of particle physics, our current best theory of the fundamental particles and forces that make up the universe. The Standard Model incorporates many key insights into the Dirac equation, including the existence of antimatter and the nature of spin.

The impact of the Dirac equation on particle physics is a testament to the power of the human mind to create something beautiful and enduring, something that captures the essence of the world in a way that transcends time and place.

5.4: Quantum Field Theory - Basics for Beginners

From Particles to Fields

Quantum field theory (QFT) combines quantum mechanics and special relativity to describe particles and fields at a fundamental level. It represents a major paradigm shift from the traditional view of particles as discrete entities moving through space to a more holistic view of particles as excitations or ripples in underlying quantum fields.

In QFT, every type of particle is associated with a corresponding field that permeates all of space-time. Particles are localized excitations or quantum states of these fields, similar to ripples, which are localized disturbances on the surface of a pond. The properties of particles, such as their mass, charge, and spin, are determined by the properties of the corresponding fields.

The transition from particles to fields in QFT is like the transition from a pixelated image to a continuous one. Each pixel is a discrete entity in a pixelated image with its own color and brightness. But as the image's resolution increases, the pixels blend into a smooth, continuous picture. Similarly, in QFT, the discrete particles of quantum mechanics blend into a continuous fabric of fields, revealing a more fundamental level of reality.

Key Concepts of Quantum Field Theory

QFT is a rich and complex theory with many important concepts and techniques. Some of the key ideas include:

1. *Lagrangian formalism:* QFT is often formulated in terms of a Lagrangian, a mathematical function that describes the dynamics of the fields. The Lagrangian encodes the kinetic and potential energy of the fields and determines the equations of motion that govern their behavior.
2. *Feynman diagrams:* Feynman diagrams are a powerful tool for visualizing and calculating particle interactions in QFT. They represent particles as lines and interactions as vertices, allowing complex processes to be broken down into simpler, more manageable pieces.
3. *Renormalization:* QFT calculations often lead to infinities or divergences that must be carefully handled through renormalization. Renormalization involves redefining the theory's parameters, such as mass and charge, in a way that absorbs the infinities and yields finite, physically meaningful results.
4. *Symmetries and conservation laws:* QFT is built on the foundation of symmetries and conservation laws, such as energy conservation, momentum and charge. These symmetries play a crucial role in

determining the form of the interactions between particles and the properties of the fields.

Just as a painter uses different colors, textures, and techniques to create a cohesive and beautiful image, a quantum field theorist uses Lagrangians, Feynman diagrams, renormalization, and symmetries to create a powerful and predictive theory of the fundamental workings of the universe.

Interactions and Particles

One of the most critical aspects of QFT is its description of particle interactions. In QFT, particles interact by exchanging other particles, known as virtual particles or force carriers. For example, the electromagnetic interaction between two charged particles is mediated by the exchange of virtual photons, while the exchange of virtual gluons mediates the strong nuclear interaction between quarks.

The properties of the corresponding force carriers determine the strength and range of the interactions. Photons, for example, are massless, which allows the electromagnetic interaction to have an infinite range. Conversely, Gluons are massless but carry a type of charge called color charge, which causes the solid nuclear interaction to be short-ranged and confining.

The description of interactions and particles in QFT is like a cosmic dance, with particles twirling and swaying to the rhythms of the fundamental forces. Each interaction is like a different dance style with unique steps and patterns. And just as dancers can switch partners and styles throughout a performance, particles in QFT can change their identities and properties by exchanging virtual particles, creating a dynamic and ever-changing tapestry of interactions.

Real-World Applications

QFT is not just an abstract theory but has many critical real-world applications. Some of the most notable examples include:

1. ***Particle physics:*** QFT serves as the fundamental language of particle physics, providing a comprehensive framework for comprehending the properties and interactions of the fundamental particles that constitute the universe. It has facilitated the development of the Standard Model of particle physics, which has accurately predicted and explained a diverse range of experimental results.

2. ***Condensed matter physics:*** QFT has also found critical applications in condensed matter physics, where it describes the behavior of many-body

systems, such as superconductors and quantum liquids. In particular, the concept of quasiparticles, which are collective excitations of many-body systems that behave like particles, has been a powerful tool for understanding the properties of these systems.

3. ***Cosmology:*** QFT has been instrumental in helping us understand the early universe and how structure emerged in the cosmos. Specifically, the concept of cosmic inflation, which suggests that the universe underwent a brief but intense expansion during its initial stages, is rooted in the principles of QFT.

The real-world applications of QFT are like the branches of a great tree, each reaching out in a different direction, drawing nourishment from the same deep roots. Just as a tree can provide shade, shelter, and sustenance to various creatures, QFT delivers a framework for understanding and exploring a wide range of physical phenomena, from the tiniest scales of particle physics to the greatest scales of cosmology.

5.5: The Standard Model of Particle Physics Simplified

Building Blocks of the Universe

The Standard Model is a theory that defines the universe's fundamental components and interactions. It is established on the principles of quantum field theory and has been exceedingly successful in foretelling and interpreting a wide range of experimental results.

At the heart of the Standard Model are two types of particles: bosons and fermions. The particles that make up matter are fermions, such as electrons, quarks and neutrinos. Fermions have half-integer spin and obey the Pauli exclusion principle, which states no identical fermions can simultaneously occupy the same quantum state.

Bosons, on the other hand, are the particles that mediate the interactions between fermions. They have integer spin and do not obey the Pauli exclusion principle, allowing them to occupy the same quantum state. The most familiar boson is the photon, the light particle and the electromagnetic interaction mediator.

The Standard Model of the universe consists of fundamental particles that can be compared to the elements of a periodic table, each with its unique properties and purpose. Like the periodic table, the Standard Model organizes these particles based on their quantum numbers, such as spin and charge, revealing symmetries and patterns that form the very structure of reality.

Unification of Forces

One of the great triumphs of the Standard Model is its unification of three of the four fundamental forces of nature: the electromagnetic force, the weak nuclear force, and the strong nuclear force. The Standard Model shows that these forces are not separate entities but are different aspects of a single underlying force, known as the electroweak force.

Sheldon Glashow, Abdus Salam and Steven Weinberg first proposed the unification of the electromagnetic and weak forces in the 1960s. This was later confirmed experimentally with the discovery of the W and Z bosons, which are the mediators of the weak force. The strong force, which is responsible for holding quarks together inside protons and neutrons, is described by a separate theory called quantum chromodynamics (QCD).

Just as the waters of various rivers merge as they flow towards the sea, the other forces of nature merge at high energies, revealing their underlying unity and symmetry. The Standard Model provides a map of this cosmic watershed, showing us how the different forces are connected and how they shape the landscape of the universe.

Successes and Mysteries

The Standard Model is a scientific theory that has been immensely successful in explaining and predicting a wide range of experimental results in particle physics. It has led to groundbreaking discoveries such as the Higgs boson and allowed us to make precise measurements of fundamental particles like electrons and muons. Moreover, the development of new technologies, like the Large Hadron Collider, has enabled us to push the boundaries of our understanding of the universe to new frontiers.

However, the Standard Model is only a partial theory of everything. It does not include gravity, described by Einstein's theory of general relativity, and it does not explain some of the universe's most profound mysteries, such as the nature of dark matter and dark energy, which make up most of the universe's mass and energy.

Moreover, the Standard Model has several parameters, such as the particles' masses and the interactions' strengths, that are not predicted by the theory but must be measured experimentally. This has led some physicists to search for a more fundamental theory that can explain these parameters and provide a more complete understanding of the universe.

Beyond the Standard Model

Despite its many successes, the Standard Model is not the final word on the nature of the universe. There are many open questions and challenges that remain, and physicists are actively searching for new theories and ideas that can take us beyond the Standard Model.

One of the most promising areas of scientific exploration is supersymmetry, which proposes that each particle in the Standard Model has a supersymmetric partner with different properties. Supersymmetry could help to solve some of the Standard Model's mysteries, such as the hierarchy problem of why the Higgs boson is so much lighter than the other particles. Additionally, it could provide a solution for the elusive dark matter.

Another exciting frontier in physics is the search for a theory of quantum gravity. Such a theory would unify the principles of quantum mechanics and general relativity into a single, coherent framework. String theory and loop quantum gravity are two of the most promising contenders for this theory. However, many challenges and open questions still need to be addressed in order to achieve this goal.

5.6: Gravitational Quantum Mechanics

The Challenge of Quantum Gravity

One of the most significant challenges in modern physics is the quest for a theory of quantum gravity, which would unite the principles of quantum mechanics and general relativity into a single, coherent framework. The need for such a theory arises from the fact that both quantum mechanics and general relativity are incomplete on their own, and they break down in situations where both theories are needed, such as in the early universe or near the center of a black hole.

The challenge of quantum gravity arises because the principles and mathematical frameworks of quantum mechanics and general relativity are very different. Quantum mechanics explains the behavior of particles at the microscopic level, where the influence of gravity is insignificant. On the other hand, general relativity describes the behavior of space and time at the macroscopic level, where the influence of gravity is dominant.

Attempting to combine these two theories leads to several conceptual and mathematical difficulties. For example, in quantum mechanics, particles are described by wave functions simultaneously in multiple states. In contrast, in general relativity, particles follow definite paths through space and time. Similarly, in quantum mechanics, the uncertainty principle limits the precision with which specific properties of particles can be measured. In contrast, in general relativity, the properties of space and time are determined with absolute precision.

Current Approaches

Physicists have made significant progress in the quest for a theory of quantum gravity despite facing challenges. Multiple approaches with varying strengths and weaknesses are currently being pursued.

One of the most promising approaches is string theory, which proposes that the universe's fundamental building blocks are not point-like particles but rather one-dimensional strings that vibrate differently. String theory has the potential to unify all the forces of nature, including gravity, into a single, coherent framework. Still, it has some unresolved issues, such as possible solutions and difficulty making testable predictions.

Another approach is loop quantum gravity, which quantizes space and time by breaking them down into discrete chunks or loops. Loop quantum gravity has had some success in resolving the singularities that arise in classical general relativity, such as the Big Bang and the centers of black holes. However, it is still a work in progress, and many open questions remain.

Other approaches include causal dynamical triangulations, which model space and time as a network of evolving triangles, and the holographic principle, which suggests that the information in a region of space can be described by a theory that lives on the boundary of that region.

Experimental Challenges

One of the biggest challenges in the quest for quantum gravity is the difficulty of testing the theories experimentally. The energy scales at which quantum gravity effects become significant are beyond current particle accelerators, making it difficult to probe space and time at the smallest scales.

Moreover, the effects of quantum gravity are expected to be extremely weak at the scales accessible to current experiments. For example, the gravitational force between two protons is roughly 10^{-36} times weaker than the electromagnetic force, making it incredibly difficult to detect.

Despite these challenges, physicists are developing new experimental techniques and technologies that could help test the predictions of quantum gravity theories. Some researchers, for example, are exploring the possibility of using ultra-precise atomic clocks to detect tiny variations in the flow of time that quantum gravity effects could cause. Others are looking for ways to use gravitational wave detectors to probe the properties of space and time at the smallest scales.

Physicists working on quantum gravity must use cutting-edge technologies and innovative approaches to probe the properties of space and time at the smallest scales. It is a daunting task, but one that holds the promise of revealing the hidden wonders of the universe.

Implications for Cosmology

A theory of quantum gravity would have profound implications for our understanding of the universe. It could help to explain some of the deepest mysteries of cosmology, such as the nature of dark matter and dark energy, the origin of the universe, and the ultimate fate of black holes.

For example, some theories of quantum gravity, such as loop quantum gravity, suggest that the Big Bang was not the beginning of time but rather a transition from a previous state of the universe. Other theories, such as string theory, suggest that our universe may be just one of many in a vast multiverse, each with its unique properties and laws of physics.

A theory of quantum gravity could also help resolve some paradoxes that arise in classical general relativity, such as the information paradox of black holes. According to classical general relativity, anything that falls into a black hole is lost forever, including all the information it contains. However, quantum mechanics suggests that information cannot be lost, leading to a paradox. A quantum gravity theory could resolve this paradox by describing how information is preserved in a quantum mechanical description of black holes.

Conclusion

As we have seen, searching for a theory of quantum gravity is one of the most significant challenges in modern physics, with profound implications for our understanding of the universe. It requires us to confront some of the most profound questions about the nature of space, time, and matter and push the boundaries of our current theories and experimental techniques.

Despite the challenges, physicists are progressing on many fronts, from developing new theoretical frameworks to designing cutting-edge experimental technologies. While a complete theory of quantum gravity remains elusive, the insights and discoveries made along the way will deepen our understanding of the universe and our place within it.

As we embark on a journey of exploration, it is crucial to remember that pursuing knowledge is not only a scientific endeavor but also a profoundly human one. This quest has driven us to explore the farthest reaches of the universe and the deepest depths of the subatomic world, and it will continue to inspire us as we strive to unravel the mysteries of the cosmos.

Ultimately, the search for quantum gravity is a testament to the power of the human mind to ask profound questions and seek answers, no matter how difficult or elusive they may be.

6

Quantum Paradoxes and Thought Experiments

Chapter 6: Quantum Paradoxes and Thought Experiments

In this chapter, we will explore some of the most famous and perplexing thought experiments in the history of quantum mechanics, from Schrödinger's Cat to the EPR Paradox and from the Many-Worlds Interpretation to the Delayed Choice Quantum Eraser. Along the way, we will explore the role of the observer, the nature of measurement, and the implications of these experiments for our understanding of reality.

Get ready for an incredible journey through the world of quantum mechanics, where the impossible becomes possible, and the unimaginable becomes a reality!

6.1: Schrödinger's Cat and the Observer's Role

Illustrating Superposition

In a hypothetical scenario, a cat is sealed in a box with poison and a radioactive source. The cat is both dead and alive until the box is opened, according to quantum mechanics. Opening the box collapses the wave function, forcing the cat into one state. This is the essence of Schrödinger's Cat, a thought experiment that has captivated the minds of scientists and philosophers alike for nearly a century.

At the heart of this thought experiment lies superposition. Superposition means that, until it is observed or measured, a quantum system exists simultaneously in multiple states. In the case of Schrödinger's Cat, the radioactive source is coupled to a device that will shatter the vial of poison if a single atom decays. Since the atom's decay is a quantum event governed by probability, the cat's fate is entangled with the atom's state, and it exists in a superposition of alive and dead states until the box is opened.

The idea of a cat being alive and dead simultaneously may seem absurd, but it is a direct consequence of quantum mechanics' mathematical formalism. The Schrödinger equation governs quantum systems' evolution and allows for superposition states, where multiple outcomes coexist in a single system. Only when we observe the system does the wave function collapse, forcing it into a definite state.

Debate on Observation

The role of the observer in collapsing the wave function is a topic of much debate and controversy in the quantum physics community. Some interpretations, such as the Copenhagen Interpretation, suggest that observation causes the wave function to collapse. This means that the observer plays a critical role in determining the outcome of a quantum measurement.

Other interpretations, such as the Many-Worlds Interpretation, suggest that the wave function never actually collapses but that all possible measurement outcomes co-occur in different branches of reality. In this view, the observer does not play a unique role in determining the outcome but instead experiences only one branch of the many possible realities.

The debate over the role of the observer in quantum mechanics has led to some fascinating philosophical questions about the nature of reality itself. If the act of observation can fundamentally alter the state of a quantum system, then what does that say about the objectivity of reality? Are we, as observers, active participants in shaping the fabric of the universe, or are we merely passive witnesses to a predetermined unfolding of events?

Macroscopic vs. Microscopic

One key challenge in interpreting Schrödinger's Cat is how quantum phenomena, typically observed at the microscopic scale, can be applied to macroscopic objects like cats. In the everyday world, we don't see objects existing in superposition states or being affected by the act of observation. So, why should a cat be any different?

The answer lies in Decoherence, which describes the interaction of quantum systems with their environment and the loss of their coherence over time. In the case of Schrödinger's Cat, the cat is not a truly isolated quantum system. Still, it constantly interacts with its environment, including the air in the box, the walls of the box, and even the atoms in its own body. These interactions cause the superposition state to decohere rapidly, effectively forcing the cat into a definite state long before opening the box.

The fact that we don't observe quantum phenomena in the macroscopic world is not a limitation of quantum mechanics but rather a consequence of the complex interactions between quantum systems and their environments. Understanding the concept of Decoherence is critical in explaining how the quantum world transforms into the classical world that we can observe. Additionally, it has significant implications for advancing quantum technologies like quantum computers and sensors.

Implications for Reality

The implications of Schrödinger's Cat for our understanding of reality are profound and far-reaching. If the act of observation can fundamentally alter the state of a quantum system, then what does that say about the nature of reality itself? Is reality an objective, pre-existing entity we merely observe, or is it a participatory process actively shaped by our interactions?

The Copenhagen Interpretation suggests that reality is not a fixed, objective entity but a fluid and malleable construct constantly being shaped by our observations and measurements. In this view, the observer is not a passive witness to reality but an active participant in its creation.

The idea that reality is participatory and observer-dependent has deep philosophical and spiritual implications. It suggests that consciousness itself plays a fundamental role in the fabric of the universe and that our thoughts, beliefs, and intentions directly impact the world around us. This idea has been explored in various contexts, from the Buddhist concept of dependent origination to the philosophy of idealism, and it continues to inspire new ways of thinking about the nature of reality and our place within it.

6.2: The EPR Paradox - Questioning Locality and Reality

Einstein's Discomfort

In the early days of quantum mechanics, only some were convinced by the theory's strange and counterintuitive predictions. One of the most famous skeptics was Albert Einstein, who famously declared that "God does not play dice with the universe." Einstein's discomfort with quantum mechanics stemmed from his belief in a deterministic and objective reality, where the outcomes of measurements are predetermined by the underlying properties of the system rather than being fundamentally probabilistic.

In 1935, Einstein, alongside his colleagues Boris Podolsky and Nathan Rosen, published a paper that challenged the completeness of quantum mechanics. They argued for hidden variables that could explain the apparent randomness of quantum measurements. This argument, known as the EPR Paradox, was based on entanglement. Entanglement allows two particles to be correlated so that measuring one particle immediately determines the state of the other, even if the particles are far apart.

Einstein's discomfort with quantum mechanics was not just a matter of personal philosophy but a deep-seated belief in the fundamental principles of locality and realism. Locality is the idea that the outcome of a measurement of one particle cannot be influenced by the measurement of another particle unless there is a direct causal connection between them. Realism is the idea that the properties of a system exist objectively, independent of any measurement or observation. The EPR Paradox challenged both of these principles, suggesting that quantum mechanics was incomplete or violated the basic tenets of special relativity.

Nonlocality and Entanglement

The EPR Paradox is based on the concept of quantum entanglement. It is a phenomenon where two or more particles are interrelated in that the state of one particle is dependent on the state of the others. This correlation is unique to the quantum world and has no classical equivalent. Entanglement is a fundamental concept in quantum mechanics responsible for its confusing and counterintuitive properties.

The EPR Paradox considers a pair of entangled particles, such as two electrons with opposite spins. According to quantum mechanics, the spin of each electron is in a superposition state of up and down until it is measured, at which point it collapses into a definite state. However, if the two electrons are entangled, measuring one electron's spin instantaneously determines the other's spin, regardless of distance.

This apparent nonlocality, where the state of one particle can be influenced by the measurement of another particle across vast distances, challenged Einstein's belief in the principle of locality and suggested that quantum mechanics was incomplete. Einstein argued that hidden variables must determine the outcome of the measurement and that the apparent randomness of quantum mechanics reflects our ignorance of these hidden variables.

The concept of nonlocality in quantum mechanics has great implications in our understanding of reality. Suppose the state of one particle can be instantaneously influenced by the measurement of another particle, even if vast distances separate the particles. In that case, it suggests that the universe is fundamentally interconnected in ways we are only beginning to understand. Some interpretations of quantum mechanics, such as the Bohm interpretation, posit the existence of a quantum potential that permeates all of space and allows for instantaneous communication between entangled particles. Other interpretations, such as the Many-Worlds Interpretation, suggest that entanglement results from the branching of reality into multiple parallel universes. Regardless of the interpretation, the phenomenon of quantum entanglement challenges our classical notions of locality and realism and suggests that the fabric of reality is far more complex and subtle than we ever imagined.

Bell's Theorem

In 1964, physicist John Stewart Bell proposed a theorem that tested the predictions of quantum mechanics against those of local hidden variable theories, such as the one proposed by Einstein in the EPR Paradox. Bell's Theorem demonstrated that any theory based on local hidden variables that produce the same predictions as quantum mechanics must satisfy certain statistical limits called Bell's Inequalities.

On the other hand, Quantum mechanics predicts that Bell's Inequalities can be violated under certain conditions, such as when measuring the spins of entangled particles. If experiments could demonstrate a violation of Bell's Inequalities, it would prove that quantum mechanics is incompatible with any local hidden variable theory and that the apparent nonlocality of entanglement is a natural and fundamental feature of the universe.

Over the past few decades, numerous experiments have been conducted to test Bell's Theorem using entangled photons, electrons and even atoms. These experiments have consistently shown a violation of Bell's Inequalities, providing strong evidence for the nonlocal nature of quantum mechanics and ruling out the possibility of local hidden variable theories.

The experimental confirmation of Bell's Theorem has profoundly impacted our understanding of the nature of reality. It has shown that the apparent randomness of quantum mechanics is not simply a reflection of our ignorance of hidden variables but a fundamental feature of the universe itself. It has also opened up new avenues for developing quantum technologies, such as quantum cryptography and quantum computing, which rely on the nonlocal properties of entanglement to perform tasks that are impossible with classical systems.

Reconciling with Relativity

The nonlocality implied by Bell's Theorem and the experimental confirmation of entanglement conflict directly with the principles of special relativity, which prohibit faster-than-light communication and ensure the locality of cause and effect. This apparent conflict has led to much debate and speculation about the nature of space, time, and causality in the quantum world.

One possible resolution to this conflict is the idea of quantum nonlocality, which suggests that the apparent nonlocality of entanglement is not a violation of special relativity but rather a reflection of the fundamental interconnectedness of the universe at the quantum level. In this view, entangled particles are not communicating faster than light but are connected by a deeper level of reality that transcends the limitations of classical space and time.

Another possibility is that the nonlocality of entanglement results from the branching of reality into multiple parallel universes, as suggested by the Many-Worlds Interpretation of quantum mechanics. In this view, the measurement of one entangled particle does not instantaneously influence the other particle. Instead, it causes a branching of reality into multiple universes, each with a different measurement outcome.

The reconciliation of quantum nonlocality with special relativity remains an active area of research and debate in the physics community. Some theories, such as the Bohm and transactional interpretation, propose novel ways of understanding the nature of space, time and causality in the quantum world.

Other theories, such as quantum gravity and string theory, seek to unify quantum mechanics and general relativity into a comprehensive framework that can account for both entanglement's nonlocal properties and space-time curvature. Regardless of the ultimate resolution, the apparent conflict between quantum nonlocality and special relativity highlights the deep and subtle mysteries that still surround our understanding of the fundamental nature of reality.

6.3: The Many-Worlds Interpretation vs. Copenhagen Interpretation

Competing Interpretations

The Many-Worlds Interpretation and the Copenhagen Interpretation are two major interpretations of quantum mechanics that significantly impact our understanding of reality and the role of the observer in quantum measurements.

Niels Bohr and Werner Heisenberg formulated the Copenhagen Interpretation during the early days of quantum mechanics. It is considered the standard interpretation of the theory. According to this interpretation, the wave function provides a complete quantum system description. The act of measurement causes the wave function to collapse into a definite state, and the Born rule determines the probabilities of different outcomes. In this interpretation, the observer plays a central role in the measurement process, and the results of measurements are fundamentally probabilistic.

The Many-Worlds Interpretation, proposed by Hugh Everett in 1957, takes a radically different approach. It suggests that the wave function never actually collapses but that every possible outcome of a quantum measurement co-occurs in different "branches" of reality. In this interpretation, the observer has no unique role, and the apparent randomness of quantum measurements simply reflects that we are only aware of one branch of reality.

The debate between the Copenhagen Interpretation and the Many-Worlds Interpretation has been ongoing for decades, with proponents of each view arguing for the merits of their preferred interpretation. Supporters of the Copenhagen Interpretation say that it provides a pragmatic and operational approach to quantum mechanics, focusing on the outcomes of measurements rather than the underlying ontology of the theory. Critics, however, argue that the Copenhagen Interpretation is philosophically unsatisfying, relying on ad hoc assumptions about the role of the observer and the nature of measurement.

Proponents of the Many-Worlds Interpretation, on the other hand, argue that it provides a more elegant and consistent description of quantum mechanics, avoiding the philosophical difficulties associated with wave function collapse and the role of the observer. They also point out that the Many-Worlds Interpretation is consistent with quantum mechanics' mathematical formalism and provides a natural explanation for the apparent randomness of quantum measurements. Critics, however, argue that the Many-Worlds Interpretation is extravagant and untestable, requiring an infinite number of parallel universes that are fundamentally inaccessible to observation.

Implications for Reality

The Copenhagen Interpretation and the Many-Worlds Interpretation provide different views on the essence of reality and the observer's role in quantum measurements. According to the Copenhagen Interpretation, reality is probabilistic, and the act of measurement plays a vital role in determining quantum events' outcome. In this interpretation, the observer is not only a mere spectator of reality but an active participant in creating it.

In the Many-Worlds Interpretation, on the other hand, reality is deterministic and objective, with all possible outcomes of quantum measurements co-occurring in different branches of reality. The observer does not play a unique role in the measurement process but is simply aware of one particular branch of reality.

The implications of these different interpretations for our understanding of reality are profound and far-reaching. If the Copenhagen Interpretation is correct, then reality is fundamentally subjective and dependent on the observer, with the act of measurement playing a crucial role in determining the outcome of quantum events. This suggests that consciousness may be a fundamental aspect of reality and that the universe is not simply a collection of objective facts but a participatory process shaped by our observations and interactions.

Suppose the Many-Worlds Interpretation is correct, on the other hand. In that case, reality is a vast and complex multiverse, with infinite parallel universes branching off from every quantum event. This suggests that our universe is just one of countless others, each with its unique history and properties. It also implies that our choices and actions have consequences that ripple across the multiverse, affecting our reality and countless others.

Critiques and Support

The Copenhagen Interpretation and the Many-Worlds Interpretation have their supporters and critics within the scientific community. Critics of the Copenhagen Interpretation believe that it could be more philosophically unsatisfying because it relies on ad hoc assumptions about the role of the observer and the nature of measurement. They also point out that wave function collapse is inherently problematic, as it violates the linearity of quantum mechanics and introduces an element of nonlocality into the theory.

On the other hand, supporters of the Copenhagen Interpretation argue that it provides a pragmatic and operational approach to quantum mechanics, focusing on the outcomes of measurements rather than the underlying ontology of the theory. They also point out that the Copenhagen Interpretation has predicted a wide range of experimental results successfully and provides a natural framework for understanding the observer's role in quantum measurements.

Critics of the Many-Worlds Interpretation argue that it is extravagant and untestable, requiring an infinite number of parallel universes that are fundamentally inaccessible to observation. They also point out that the idea of parallel universes is inherently speculative, and there is no way to verify the existence of other branches of reality experimentally.

Supporters of the Many-Worlds Interpretation, on the other hand, argue that it provides a more elegant and consistent description of quantum mechanics, avoiding the philosophical difficulties associated with wave function collapse and the role of the observer. They also point out that the Many-Worlds Interpretation is consistent with quantum mechanics' mathematical formalism and provides a natural explanation for the apparent randomness of quantum measurements.

The debate between the Copenhagen Interpretation and the Many-Worlds Interpretation will likely continue for the foreseeable future, as both interpretations have their strengths and weaknesses. Some scientists have proposed alternative interpretations, such as the Consistent Histories Interpretation and the Objective Collapse Interpretation, which attempt to reconcile the advantages of the Copenhagen and Many-Worlds Interpretations while avoiding their drawbacks. Ultimately, the choice of interpretation may come down to personal philosophical preferences and the specific scientific questions and applications being considered.

Philosophical Considerations

The debate between the Copenhagen Interpretation and the Many-Worlds Interpretation is not just a scientific one but also a deeply philosophical one. At its core, this debate is about the nature of reality itself, and the role that consciousness and observation play in shaping it.

The Copenhagen Interpretation can be viewed as a form of subjective idealism from a philosophical standpoint. This perspective suggests that reality is essentially contingent on the observer and that the observation process plays an important role in establishing the nature of reality. This view is consistent with particular strands of Eastern philosophy, such as the Buddhist concept of emptiness, which holds that all phenomena are fundamentally interdependent and lack inherent existence.

On the other hand, the Many-Worlds Interpretation can be seen as a form of objective realism. It holds that reality exists independently of the observer and that the universe is a vast and complex multiverse, with infinite parallel universes branching off from every quantum event. This view is consistent with particular strands of Western philosophy, such as the materialist and determinist philosophies of the Enlightenment.

Quantum mechanics' philosophical implications have been a subject of intense debate for nearly a century. Some philosophers, such as Niels Bohr and Werner

Heisenberg, have argued that quantum mechanics requires a fundamental revision of our understanding of reality and that the classical concepts of objectivity, causality, and determinism must be abandoned in the quantum realm. Others, such as Albert Einstein and Erwin Schrödinger, have argued that quantum mechanics is incomplete and that there must be a deeper level of reality that is not yet understood.

Some philosophers have argued that quantum mechanics may have implications for the nature of consciousness itself. The philosopher David Chalmers, for example, has suggested that the Many-Worlds Interpretation of quantum mechanics may provide a natural framework for understanding the relationship between the physical world and conscious experience. According to this view, consciousness may be a fundamental aspect of reality, and the branching of the multiverse in the Many-Worlds Interpretation may correspond to the branching of conscious experience.

6.4: Quantum Eraser Experiment

Experiment Overview

The quantum eraser experiment is a variation of the double-slit experiment demonstrating quantum mechanics' strange and counterintuitive nature. In this experiment, a beam of photons is sent through a double slit, creating an interference pattern on a screen behind the slits. However, the experiment is set up in such a way that the path taken by each photon can be determined, either by placing detectors at the slits or by using polarizing filters to "tag" the photons.

According to classical physics, the presence of detectors or filters should not affect the interference pattern, as the photons are assumed to have a definite path through one slit or the other. However, in the quantum eraser experiment, the presence of detectors or filters destroys the interference pattern, as if the photons are "aware" of being observed and change their behavior accordingly.

The quantum eraser experiment takes this a step further by allowing the experimenter to "erase" the path information after the photons have already passed through the slits. This is done using a particular type of polarizing filter that can be rotated to preserve or erase the path information. When the path information is erased, the interference pattern reappears, even though the photons have already passed through the slits.

The quantum eraser experiment powerfully demonstrates the role of observation and measurement in quantum mechanics. It shows that the act of observing a quantum system can fundamentally change its behavior and that the effects of observation can persist even after the measurement has been made. This has led some scientists to speculate that the quantum eraser experiment may have implications for our understanding of causality and the nature of time itself.

Questioning Causality

The results of the quantum eraser experiment challenge our conventional notions of causality and the flow of time. In classical physics, causality is assumed to be a one-way street, with causes always preceding their effects in time. However, in the quantum eraser experiment, the decision to erase the path information can be made after the photons have already passed through the slits, suggesting that the effect (the presence or absence of an interference pattern) can precede its cause (the decision to erase the path information).

This apparent causality violation has led some scientists to speculate that quantum mechanics may allow for a form of "retrocausality," where the future can influence the past. According to this view, the decision to erase the path information in the quantum eraser experiment may actually "reach back in time" to influence the behavior of the photons as they pass through the slits.

The idea of retrocausality in quantum mechanics is highly controversial, and there needs to be a consensus among scientists about whether it is a viable explanation for the results of the quantum eraser experiment. Some have argued that the apparent violation of causality is simply a result of our limited understanding of quantum mechanics, and that a more complete theory may restore the conventional notion of causality. Others have suggested that the quantum eraser experiment requires a fundamental revision of our understanding of space and time and that the concepts of causality and the flow of time need to be re-examined in the quantum realm.

Interpretations and Implications

Different scientists and philosophers have interpreted the quantum eraser experiment in various ways. Some have argued that it demonstrates the inherently probabilistic nature of quantum mechanics and that the absence or presence of an interference pattern is simply a reflection of the probabilities associated with different measurement outcomes.

Others have suggested that the quantum eraser experiment may have implications for the nature of reality itself. According to this view, the results of the experiment suggest that reality is not a fixed and objective entity but rather a fluid and malleable construct shaped by our observations and interactions. This interpretation is consistent with certain strands of Eastern philosophy, such as the Buddhist concept of emptiness, which holds that all phenomena are fundamentally interdependent and lack inherent existence.

Scientists and philosophers are still debating and exploring the implications of the quantum eraser experiment for our understanding of reality. Some have argued that the experiment may have implications for the nature of consciousness itself and that the observer's role in quantum mechanics may be related to the nature of subjective experience. Others have suggested that the quantum eraser experiment may have practical applications in fields such as quantum computing and cryptography, where the ability to manipulate and control quantum systems is essential.

Understanding Quantum Information

At the heart of the quantum eraser experiment is the concept of quantum information, which refers to the information encoded in a system's quantum state. In the case of the quantum eraser experiment, the quantum information is contained in the polarization of the photons, which can be used to determine the path taken by each photon through the double slit.

The quantum eraser experiment demonstrates the fundamental role that quantum information plays in the behavior of quantum systems. It shows that the presence or absence of quantum information can profoundly affect the outcome of a quantum measurement, and that erasing quantum information can actually change a quantum system's behavior.

The concept of quantum information is central to many areas of quantum physics, including quantum computing, quantum cryptography, and quantum communication. In these fields, manipulating and controlling quantum information is essential for developing new technologies and applications. For example, quantum computing perform complex calculations by using quantum information, and that would be impossible with classical computers. In contrast, quantum cryptography uses quantum information to create unbreakable codes and secure communication channels.

6.5: Wigner's Friend

Thought Experiment Explained

Physicist Eugene Wigner proposed in 1961 a thought experiment called Wigner's Friend. The experiment explores the role of observation and measurement in quantum mechanics and involves two observers: Wigner and his friend. Wigner's friend performs a quantum measurement on a system in a superposition of two states, such as a photon that can be polarized horizontally or vertically. According to quantum mechanics, measurement causes the system to collapse into one of the two states, with a probability given by the Born rule.

However, Wigner himself does not directly observe the measurement but instead asks his friend about the result. From Wigner's perspective, the system is still in a superposition of states until he learns the measurement result from his friend. This creates a paradox, as it suggests that the system's state is subjective and depends on the observer.

The Wigner's Friend thought experiment highlights the role of subjectivity in quantum mechanics and raises questions about the nature of reality itself. If the state of a quantum system depends on the observer, then what does it mean for something to be "real"? Are there multiple realities that exist simultaneously, each dependent on a different observer? Or is there a single objective reality that exists independently of any observer?

Observation and Reality

The Wigner's Friend thought experiment challenges our conventional notions of objectivity and reality. In classical physics, reality is assumed to be a fixed and objective entity that exists independently of any observer. However, in quantum mechanics, the act of observation and measurement plays a crucial role in determining the state of a system, suggesting that reality may be subjective and dependent on the observer.

This idea is closely related to the concept of quantum entanglement, which describes a situation where two or more particles are correlated in such m that the state of one particle cannot be described independently of the state of the others. In the case of Wigner's Friend, the system being measured and the friend performing the measurement can be considered an entangled pair, with the state of the system dependent on the state of the friend's knowledge.

The idea of a subjective and observer-dependent reality has far-reaching implications for our understanding of the universe's nature.

If reality is not a fixed and objective entity, but rather a fluid and malleable construct that is shaped by our observations and interactions, then what does it mean for something to be "true" or "false"? Are there multiple truths that exist simultaneously, each dependent on a different observer? Or is there a single objective truth that exists independently of any observer, but is fundamentally unknowable?

Consciousness and Measurement

The Wigner's Friend thought experiment also raises questions about the consciousness' role in quantum mechanics. If the state of a quantum system depends on the observer, what exactly constitutes an observer? Is it necessary for the observer to be conscious, or can any physical system interacting with the quantum system be considered an observer?

Some interpretations of quantum mechanics, such as the von Neumann-Wigner interpretation, suggest that consciousness may play a fundamental role in the measurement process. This viewpoint suggests that the wave function collapses, and the system takes on a definite state due to conscious observation. This idea is closely related to the "hard problem of consciousness," which is the difficulty of explaining how subjective experience arises from physical processes. However, the idea of a fundamental link between consciousness and quantum mechanics remains highly speculative and controversial.

There has yet to be a consensus among scientists about whether it is a viable explanation for the measurement problem in quantum mechanics. Some have argued that a conscious observer is unnecessary and that any physical system interacting with the quantum system can be considered an observer. Others have suggested that the measurement problem may require a more radical revision of our understanding of reality, such as the Many-Worlds Interpretation or the Objective Collapse Interpretation.

Debates and Discussions

The Wigner's Friend thought experiment has been debated by scientists and philosophers since its proposal. Some have argued that the thought experiment demonstrates reality's inherently subjective nature and consciousness's fundamental role in the universe. Others have suggested that the thought experiment reflects our limited understanding of quantum mechanics and that a more complete theory may resolve the paradox without needing a conscious observer.

One of the most famous debates surrounding the Wigner's Friend thought experiment is the Frauchiger-Renner paradox, proposed by physicists Daniela Frauchiger and Renato Renner in 2018. The paradox involves a variation of the Wigner's Friend thought experiment with multiple observers and measurements, and it suggests that quantum mechanics may be inconsistent with certain assumptions about the nature of reality.

The debate surrounding the Wigner's Friend thought experiment and the Frauchiger-Renner paradox is ongoing, and there needs to be more consensus among scientists about the implications of these ideas. Some have argued that the paradoxes require a fundamental revision of quantum mechanics, such as modifying the Born rule or introducing new physical principles. Others have suggested that the paradoxes may be resolved by a more complete understanding of quantum mechanics, such as the Consistent Histories Interpretation or the Objective Collapse Interpretation.

Ultimately, the Wigner's Friend thought experiment and the debates surrounding it highlight the deep and unresolved questions at the heart of quantum mechanics. These questions touch on some of the most fundamental aspects of reality, including the nature of consciousness, the role of observation and measurement, and the relationship between the subjective and the objective.

Although we don't have all the answers yet, exploring these ideas will deepen our understanding of the universe and our place in it.

6.6: The Delayed Choice Quantum Eraser

Experiment Setup and Results

The delayed choice quantum eraser experiment is a variation of the double-slit experiment that challenges our conventional notions of causality and the flow of time. In this experiment, a beam of photons is sent through a double slit, creating an interference pattern on a screen behind the slits. However, the experiment is set up in such a way that the path taken by each photon can be determined after the photon has already passed through the slits using a device called a quantum eraser.

The quantum eraser works by entangling each photon with another, called an idler photon, sent through a separate path. By measuring the idler photon, the experimenter can determine which path the original photon took through the double slit, even after the original photon has already been detected on the screen.

The results of the delayed choice quantum eraser experiment are surprising and counterintuitive. When the path information is erased, the interference pattern reappears, even though the decision to erase the path information was made after the original photon had already been detected. This suggests that the behavior of the original photon is influenced by a decision made in the future, challenging our conventional notions of causality and the flow of time.

The delayed choice quantum eraser experiment powerfully demonstrates quantum mechanics' strange and counterintuitive nature. It shows that the behavior of quantum systems can be influenced by decisions made in the future, suggesting that the flow of time may be more complex than we typically assume. Some scientists speculate that quantum mechanics may require a fundamental revision of our understanding of causality and the nature of time itself.

Challenging Classical Intuition

Our classical intuition about causality and the flow of time are challenged by the results of the delayed choice quantum eraser experiment. In classical physics, causality is assumed to be a one-way street, with causes always preceding their effects in time. However, in the delayed choice quantum eraser experiment, the decision to erase the path information can be made after the original photon has already been detected, suggesting that the effect (the presence or absence of an interference pattern) can precede its cause (the decision to erase the path information).

This apparent causality violation has led some scientists to speculate that quantum mechanics may allow for a form of "retrocausality," where the future can influence the past. According to this view, the decision to erase the path information in the delayed choice quantum eraser experiment may actually "reach back in time" to influence the behavior of the original photon as it passes through the double slit.

The idea of retrocausality in quantum mechanics is highly controversial, and there is no consensus among scientists about whether it is a viable explanation for the results of the delayed choice quantum eraser experiment. Some have argued that the apparent violation of causality is simply a result of our limited understanding of quantum mechanics and that a more complete theory may restore the conventional notion of causality. Others have suggested that the delayed choice quantum eraser experiment requires a fundamental revision of our understanding of space and time, and that the concepts of causality and the flow of time must be re-examined in the quantum realm.

Implications for Quantum Theory

The delayed choice quantum eraser experiment is significant as it sheds light on our comprehension of quantum mechanics and the fundamental nature of reality. Its findings suggest that the behavior of quantum systems is not solely dependent on their initial conditions and the forces acting upon them but also on the choices made by observers and the information they have available to them.

The concept of quantum entanglement is closely related to the idea of this theory. It refers to a situation where two or more particles are correlated so that the state of one particle cannot be described independently of the state of the others. In the delayed choice quantum eraser experiment, we can think of the original photon and the idler photon as an entangled pair. The state of the original photon is dependent on the information available about the idler photon.

Scientists and philosophers are still debating and exploring the implications of the delayed choice quantum eraser experiment for quantum theory. Some argue that the experiment demonstrates the probabilistic nature of quantum mechanics and that the presence or absence of an interference pattern reflects the probabilities associated with different measurement outcomes. Others suggest that the experiment may have implications for the nature of reality itself and that the behavior of quantum systems may be influenced by the choices and actions of observers in ways that are not yet fully understood.

The Role of Information

At the heart of the delayed choice quantum eraser experiment is the role of information in quantum mechanics. The experiment demonstrates that the information available about a quantum system can profoundly affect its behavior, even if that information is obtained after the system has already been measured.

This idea is closely related to the concept of quantum information, which refers to information encoded in a system's quantum state. In the delayed choice quantum eraser experiment, the quantum information is contained in the entanglement between the original photon and the idler photon, and the option to erase or preserve this information can have a dramatic effect on the behavior of the original photon.

The role of information in quantum mechanics is a subject of active research and debate among scientists and philosophers. Some have argued that information may be the fundamental building block of reality and that the behavior of quantum systems may be determined by the flow and processing of information at the most basic level. Others have suggested that the role of information in quantum mechanics may reflect our limited understanding of the theory and that a more complete understanding may require a different conceptual framework altogether.

The delayed choice quantum eraser experiment reveals the significant role of information in quantum mechanics. It shows how our choices and actions can have unexpected and counterintuitive effects on quantum systems. This experiment invites us to question our assumptions about the nature of reality and the connection between the observer and the observed. It encourages us to explore the mysteries and marvels of the quantum realm with an inquisitive and open mind.

Conclusion

Quantum paradoxes and thought experiments, such as Schrödinger's Cat, the EPR Paradox, the Many-Worlds Interpretation, and the Delayed Choice Quantum Eraser, challenge our conventional notions of reality, causality, and observation. These ideas push the boundaries of our understanding and highlight the crucial role of the observer in quantum mechanics.

The realization that observation and measurement shape the behavior of quantum systems has profound implications for our understanding of reality, suggesting that it is a fluid and malleable construct. It challenges us to re-examine our assumptions about causality, time, and the nature of information and to explore the mysteries of the quantum realm with an open mind.

As we continue to explore the frontiers of quantum mechanics, there is still much to be discovered. The paradoxes and thought experiments discussed in this chapter are just the beginning, and many more surprises await us. Whether we are probing the nature of reality, the role of consciousness, or the flow of information in the universe, quantum mechanics promises to be a rich and endlessly fascinating field of study.

Ultimately, quantum paradoxes and thought experiments remind us that the universe is far stranger and more mysterious than we might have imagined. They invite us to embrace the unknown, challenge our assumptions, and approach the world with wonder and humility as we continue to explore the quantum realm.

Quantum Mechanics for the Curious Mind

Chapter 7: Quantum Mechanics for the Curious Mind

Welcome back, dear reader, to our ongoing exploration of the fascinating world of quantum mechanics. In the previous chapters, we delved into the fundamental principles, mathematical formalism, mind-bending applications, and philosophical implications of this revolutionary theory. It's time to demystify quantum mechanics and equip you with the tools and mindset needed to further your quantum journey.

In this chapter, we will address common misconceptions about quantum mechanics, explore its portrayal in popular culture, and provide guidance on developing a quantum mindset. We will also delve into the current frontiers and future challenges of quantum research, explore the intriguing connections between quantum mechanics and consciousness, and offer a curated selection of learning resources to help you deepen your understanding of this fascinating field.

7.1: Addressing Common Misconceptions in Quantum Mechanics

Clarifying Misconceptions

As we explore quantum mechanics, we must address common misconceptions about this fascinating but often misunderstood field. One of the most pervasive myths is that quantum mechanics requires a conscious observer to collapse the wave function and determine the outcome of a measurement.

This notion, popularized by misinterpretations of the Copenhagen Interpretation and the famous Schrödinger's Cat thought experiment, suggests that conscious observation causes a quantum system to assume a definite state. However, this fundamentally misunderstands what " observation " means in quantum mechanics.

In reality, an observation in quantum mechanics refers to any interaction with the environment that causes decoherence and the collapse of the wave function. This can be achieved by any macroscopic object, such as a measurement device, and does not require the presence of a conscious observer. The key is the exchange of information between the quantum system and its environment, not the subjective experience of an observer.

Another common misconception is that quantum mechanics is inherently random and unpredictable. While it's true that the outcomes of individual quantum measurements are probabilistic, the probabilities themselves are governed by the mathematical formalism of quantum mechanics, which provides a deterministic description of the evolution of quantum systems. The apparent randomness arises from the fundamental nature of quantum systems, not from the theory's lack of predictability or determinism.

Quantum vs. Classical

To further clarify the misconceptions surrounding quantum mechanics, it's helpful to understand the critical differences between the quantum world and the classical world we're familiar with. In the macroscopic, classical world, objects have well-defined properties, such as position and momentum, and behave in a predictable, deterministic manner according to Newton's laws of motion.

In the quantum world, however, particles exhibit wave-particle duality, existing in superpositions of multiple states until measured. Properties like position and momentum are no longer well-defined but are instead described by probability distributions. The behavior of quantum systems is inherently probabilistic, and classical intuition often breaks down when trying to understand quantum phenomena.

One way to think about the difference between quantum and classical mechanics is in terms of the information available about a system. In classical mechanics, we can, in principle, know everything about a system's state and use that information to predict its future behavior with certainty. In quantum mechanics, there is a fundamental limit to how much information we can have about a system, as described by the Heisenberg Uncertainty Principle. This limitation is not due to our ignorance or measurement errors but is a fundamental property of the quantum world itself.

Misinterpretations of Experiments

Many misconceptions about quantum mechanics stem from misinterpretations of crucial experiments, such as the double-slit and the quantum eraser experiments. These experiments are often presented as evidence that consciousness affects reality or that particles can communicate instantaneously across vast distances.

However, a closer examination of these experiments reveals that they demonstrate the wave-particle duality of matter, the role of measurement in collapsing the wave function, and the fundamentally probabilistic nature of quantum systems. These experiments' apparent "weirdness" arises from the counterintuitive behavior of quantum systems, not from any mystical properties of consciousness or instantaneous communication.

Another commonly misinterpreted experiment is the EPR paradox and Bell's Inequality, which demonstrates the phenomena of quantum entanglement and nonlocality. While these experiments show that entangled particles can exhibit correlations that cannot be explained by classical physics, they do not imply that information is being transmitted faster than the speed of light or that the particles communicate instantaneously. The correlations arise from the quantum description of the entangled system as a whole, not from any physical influence or signal passing between the particles.

Media and Misrepresentation

The misrepresentation of quantum mechanics in popular media and literature has also spread misconceptions about the field. Movies, TV shows, and books often use quantum mechanics as a plot device to justify fantastical or supernatural phenomena, such as time travel, parallel universes, or psychic powers.

While these fictional portrayals can be entertaining and thought-provoking, they often bear little resemblance to the actual principles and predictions of quantum mechanics. They can create a distorted public perception of the field, leading people to associate quantum mechanics with mysticism, pseudoscience, and other unscientific ideas.

One example of the misrepresentation of quantum mechanics in popular culture is the movie "What the Bleep Do We Know!?", which presents a mix of scientific facts and New Age spiritualism, often blurring the lines between the two. The movie makes several inaccurate claims about quantum mechanics, such as the idea that consciousness can directly affect the outcome of quantum experiments or that quantum effects can explain supernatural phenomena like telekinesis and ESP. These claims are not supported by scientific evidence and contribute to the public's confusion about the nature and implications of quantum mechanics.

7.2: Quantum Mechanics in Popular

Quantum Physics in Movies and Books

The enigmatic nature of quantum mechanics has captured the imagination of writers, filmmakers and artists, leading to a wide range of portrayals in popular culture. From mind-bending sci-fi novels to Hollywood blockbusters, quantum physics has been a source of inspiration and a tool for storytelling, often blurring the lines between science and fiction.

One of the most famous examples of quantum mechanics in literature is Douglas Adams' "The Hitchhiker's Guide to the Galaxy" series. The series features a supercomputer named Deep Thought designed to answer the "Ultimate Question of Life, the Universe, and Everything." The novels playfully explore concepts like parallel universes, time travel, and the nature of reality, often drawing on (and sometimes misrepresenting) ideas from quantum physics.

In the realm of cinema, movies like "Interstellar," "Coherence," and "Quantum of Solace" have all incorporated elements of quantum mechanics into their plotlines, with varying degrees of scientific accuracy. These films often use quantum concepts like entanglement, superposition, and the many-worlds interpretation as plot devices to explore themes of love, loss, and the nature of reality.

Another notable example of quantum mechanics in popular culture is the TV show "Manifest," which follows the mysterious reappearance of a group of airline passengers who had been presumed dead for over five years. The show incorporates elements of quantum entanglement and parallel universes into its mythology, using them to explain the strange events and abilities experienced by the characters. While the show's portrayal of quantum mechanics is largely fictionalized, it demonstrates the enduring fascination with quantum concepts and their potential for storytelling.

Separating Science from Science Fiction

While incorporating quantum mechanics into popular culture can be a great way to spark interest and curiosity about the field, it's important to separate the fictional embellishments from the scientific facts. Many popular portrayals of quantum mechanics take significant liberties with the actual science, using quantum concepts as a metaphor or a plot device rather than an accurate representation of the underlying physics.

For example, the movie "What the Bleep Do We Know!?" presents a mix of scientific facts and New Age spiritualism, often blurring the lines between the two. The movie makes several inaccurate claims about quantum mechanics, such as the idea that consciousness can directly affect the outcome of quantum experiments or that quantum effects can explain supernatural phenomena like telekinesis and ESP.

Similarly, the movie "Interstellar" uses the concept of quantum entanglement to justify faster-than-light communication and time travel, both of which are not possible according to our current understanding of quantum mechanics and relativity. While these fictional elements make for an engaging story, they are not scientifically accurate and should not be taken as a representation of real quantum physics.

It's important to note that while many popular portrayals of quantum mechanics are fictitious or exaggerated, they can still serve a valuable purpose in inspiring interest and curiosity about the field. By introducing quantum concepts to a broad audience and sparking the imagination, these portrayals can encourage people to learn more about the actual science behind the stories. However, it's crucial for science communicators and educators to provide clear and accurate explanations of quantum mechanics, to help people distinguish between fact and fiction and develop a deeper understanding of the true nature of the quantum world.

Impact on Public Understanding

The portrayal of quantum mechanics in popular culture has significantly impacted public understanding and perception of the field. On the one hand, the widespread use of quantum concepts in movies, books and TV shows has helped raise awareness and interest in quantum physics, exposing many people to ideas they may not have encountered otherwise.

On the other hand, the often sensationalized and inaccurate depictions of quantum mechanics in popular culture can lead to widespread misconceptions and misunderstandings about the field's nature. Many people associate quantum mechanics with mysticism, pseudoscience, and other unscientific ideas rather than recognizing it as a rigorous and well-established scientific theory.

The impact of popular culture on public understanding of quantum mechanics can be seen in the prevalence of quantum-inspired products and services, such as quantum healing, quantum consciousness, and quantum mysticism. These practices often invoke quantum mechanics to lend scientific credibility to unproven or pseudoscientific claims, exploiting the public's fascination with quantum concepts while misrepresenting the science behind them. This can lead to skepticism and confusion about the legitimacy of quantum mechanics as a scientific field and can make it harder for people to separate fact from fiction when learning about the subject.

Educational Opportunities

Despite the challenges posed by the misrepresentation of quantum mechanics in popular culture, there are also many opportunities to use these portrayals as a tool for science education and outreach. By leveraging the public's interest in quantum concepts and the popularity of quantum-inspired media, educators and science communicators can engage people in learning about the real science behind the stories.

For example, educators can use movies and TV shows featuring quantum mechanics as a starting point for discussions about the actual science, pointing out where the portrayals are accurate and diverge from reality. They can use fictional elements to introduce actual quantum concepts, such as superposition, entanglement, and the measurement problem, and provide clear explanations of how these concepts are understood and applied in actual quantum physics research.

Science communicators can also create educational content that piggybacks on the popularity of quantum inspired media, such as videos, blogs and social media posts that explain the science behind the stories. By meeting people where they are and using engaging, accessible language and visuals, science communicators can help to bridge the gap between popular culture and scientific understanding.

One example of using popular culture as a tool for quantum physics education is the "Science of Interstellar" book and website, which physicist Kip Thorne created to accompany the movie "Interstellar." The book and website provide in-depth explanations of the scientific concepts featured in the movie, including black holes, wormholes, and the nature of time and space. By using the movie as a starting point and providing clear, accurate explanations of the underlying science, Thorne was able to engage a broad audience in learning about some of the most fascinating and complex ideas in modern physics.

7.3: How to Think Like a Quantum Physicist

Adopting a Quantum Mindset

As we delve deeper into the world of quantum mechanics, it becomes clear that understanding this fascinating field requires more than just a grasp of the mathematical formalism and experimental results. To appreciate quantum physics's complexity and beauty, one must adopt a new way of thinking. This quantum mindset challenges classical intuition and embraces the strange and counterintuitive aspects of the quantum world.

The first step in adopting a quantum mindset is to let go of the deterministic, reductionist thinking that often characterizes classical physics. In the classical world, objects have well-defined properties and behave predictably. A system's behavior can be understood by breaking it down into its constituent parts. In the quantum world, however, particles exist in superpositions of multiple states, and their properties are described by probability distributions rather than deterministic values.

To think like a quantum physicist, one must become comfortable with uncertainty and ambiguity and learn to reason regarding probabilities and statistical distributions. It requires a shift from focusing on individual particles and their trajectories to considering the quantum system as a whole and the probabilities of different measurement outcomes.

One way to cultivate a quantum mindset is to practice thinking about wave functions and probability amplitudes rather than classical trajectories and deterministic outcomes. When considering a quantum system, visualize the wave function as a complex-valued field that encodes the probabilities of different measurement outcomes. Then, consider how the wave function evolves according to the Schrödinger equation. By becoming familiar with the mathematical tools and concepts of quantum mechanics, you can develop an intuition for how quantum systems behave and how they differ from classical systems.

Critical Thinking Skills

One important aspect of adopting a quantum physicist's mindset is the skill of critical thinking and the ability to question assumptions. Quantum mechanics has many counterintuitive concepts, like particles existing in multiple states simultaneously or the measurement's role in collapsing the wave function. These ideas challenge our classical intuition and demand a willingness to question our long-held beliefs about the nature of reality.

To enhance your critical thinking abilities concerning quantum mechanics, it's important to approach the subject with an open mind and a willingness to engage with ideas that may seem strange or paradoxical at first. This means being willing to question your assumptions and biases and consider multiple perspectives and interpretations of quantum phenomena.

One helpful tool for critical thinking in quantum mechanics is thought experiments, such as the famous Schrödinger's Cat experiment or the EPR paradox. These thought experiments are designed to highlight the counterintuitive aspects of quantum mechanics and challenge our classical intuition, forcing us to confront the limitations of our everyday experience and think more deeply about the nature of reality.

Another important aspect of critical thinking in quantum mechanics is the ability to distinguish between scientific facts and speculative interpretations. The mathematical formalism of quantum mechanics is well-established and has been extensively verified by experimentation, but there are still many open questions and debates about the interpretation of quantum phenomena, such as the nature of the wave function and the role of consciousness in measurement. As a critical thinker, it's crucial to differentiate scientific facts from philosophical interpretations and analyze them skeptically.

Conceptual vs. Mathematical Understanding

One of the challenges of learning quantum mechanics is its abstract and mathematical nature, which can be intimidating for those without a strong background in advanced mathematics. While a deep understanding of mathematical formalism is essential for researchers working on the frontiers of quantum physics, it's also possible to develop a conceptual understanding of quantum mechanics without getting bogged down in the details of the equations.

To gain a proper understanding of quantum mechanics, it's important to focus on the key ideas and principles that underlie the theory, such as wave-particle duality, superposition, entanglement, and the role of measurement in collapsing the wave function. By exploring these concepts through analogies, thought experiments, and visual aids, one can start to develop an intuitive grasp of how quantum systems behave and how they differ from classical systems.

At the same time, it's important not to rely too heavily on classical analogies or intuition when trying to understand quantum mechanics. Quantum mechanics has many aspects that seem counterintuitive. For instance, particles can exist in multiple states at once, and the wave function collapses instantly upon measurement. These concepts have no classical analog and require an open-mindedness to embrace the strange and unfamiliar.

One way to balance conceptual and mathematical understanding in quantum mechanics is to use visual aids and simulations to explore the behavior of quantum systems. Many excellent online resources and software packages allow users to simulate and visualize quantum systems, such as the double-slit experiment or the quantum harmonic oscillator. By interacting with these simulations and exploring the effects of different parameters and measurement settings, one can better understand how quantum systems behave and how they are described mathematically.

Learning to Embrace Complexity

Perhaps the most important mindset shift in learning to think like a quantum physicist is the willingness to embrace complexity and resist the temptation to oversimplify or reduce quantum phenomena to classical analogies. Quantum mechanics is inherently complex and counterintuitive, and attempts to explain it in terms of everyday experience or classical physics often fall short.

To truly embrace the complexity of quantum mechanics, one must be willing to accept the discomfort and uncertainty that come with exploring a world that is fundamentally different from our everyday experience. This means being open to new ideas and perspectives and willing to engage with concepts that may seem strange or paradoxical at first.

It also means admitting when we don't fully understand something and approaching the subject with humility and curiosity. Quantum mechanics is a vast and complex field; even the most experienced researchers are still grappling with many open questions and unresolved issues.

One way to cultivate a mindset of embracing complexity is to engage with the philosophical and interpretive aspects of quantum mechanics and the purely scientific and mathematical aspects. By exploring the different interpretations of quantum phenomena, such as the many-worlds interpretation, the Copenhagen interpretation or the pilot wave theory, one can start to appreciate the depth and subtlety of the subject and the ongoing debates and questions that continue to drive research in the field. This can also foster a sense of humility and openness to new ideas, as well as a recognition of the limits of our current understanding.

7.4: The Future of Quantum Research

Current Frontiers in Quantum Physics

As we have seen throughout this book, quantum mechanics has revolutionized our understanding of our universe and led to countless technological innovations and applications. However, despite the immense progress made in the field over the past century, many open questions and challenges remain at the frontiers of quantum research.

One of the most active and promising areas of quantum research today is quantum computing and information processing. Quantum mechanics computers use quantum mechanics principles to perform calculations that are intractable for classical computers. They have the potential to revolutionize fields ranging from cryptography and drug discovery to machine learning and optimization.

However, building practical and scalable quantum computers remains challenging due to the fragility of quantum states and the difficulty of maintaining coherence and controlling errors in large-scale quantum systems. Researchers around the world are working on developing new quantum computing architectures, error correction schemes, and algorithms to overcome these challenges and realize quantum computing's full potential.

Another exciting frontier in quantum research is the study of quantum materials and superconductors. Quantum materials are materials whose properties are dominated by quantum effects, such as superconductivity, topological insulators, and quantum spin liquids. These materials exhibit exotic and often counterintuitive behaviors that challenge our classical understanding of matter and have the potential to revolutionize fields such as quantum computing, energy storage and sensing.

Researchers are also exploring the fundamental nature of space, time, and gravity at the quantum scale through the study of quantum gravity and string theory. These theories aim to merge quantum mechanics with Einstein's theory of general relativity and establish a basis for comprehending the genesis and development of the universe at its core. While these theories remain highly speculative and have yet to be experimentally verified, they offer tantalizing glimpses into the deepest mysteries of the cosmos and the nature of reality itself.

Technological Challenges

As quantum research pushes the boundaries of our understanding and technological capabilities, it also faces significant challenges in experimental realization and practical implementation. One of the biggest challenges is decoherence, which refers to the loss of quantum coherence and information due to interactions with the environment.

Decoherence is a major obstacle for quantum computing and communication, as it limits the ability to maintain and manipulate quantum states over long distances and time scales. Researchers are creating new methods for reducing decoherence, such as topological quantum computing and quantum error correction. However, these approaches are still in the early stages of development and face significant technical hurdles.

Another challenge is the development of new quantum sensing and measurement technologies, which are essential for probing the quantum world with ever-greater precision and resolution. Superconducting quantum interference devices (SQUIDs) and nitrogen-vacancy centers in diamond are quantum sensors that have the potential to revolutionize fields such as navigation, medical imaging and fundamental physics, but they require exquisite control and sensitivity at the nanoscale level.

A promising approach to overcoming these challenges is the development of hybrid quantum systems, which combine multiple quantum technologies and platforms to leverage their complementary strengths and mitigate their weaknesses. For example, researchers are exploring the integration of superconducting qubits with spin qubits in silicon or the use of photonic qubits for long-distance quantum communication and networking.

Another approach is quantum simulation and emulation, which involves using well-controlled quantum systems to simulate and study the behavior of more complex and inaccessible quantum systems. This approach has already led to significant advances in condensed matter physics and high-energy physics and has the potential to provide new insights into the fundamental workings of the quantum world.

Interdisciplinary Opportunities

As quantum research pushes into new frontiers and applications, it is also becoming increasingly interdisciplinary, drawing on expertise and insights from various fields beyond physics. This multidisciplinary approach is essential for addressing quantum technologies' complex challenges and opportunities and realizing their full potential for science and society.

One of the most exciting areas of interdisciplinary quantum research is the intersection of quantum physics with biology and medicine. Researchers are exploring the role of quantum effects in biological processes such as photosynthesis and avian navigation and developing new quantum-enhanced imaging and sensing technologies for medical diagnosis and treatment.

Another area of interdisciplinary opportunity is the intersection of quantum physics with computer science and information theory. Quantum algorithms and protocols, such as Shor's algorithm for factoring large numbers and quantum key distribution for secure communication, can revolutionize fields such as cryptography, optimization, and machine learning and drive new collaborations between physicists, computer scientists and mathematicians.

Quantum research also intersects with chemistry and materials science, as researchers seek to design and synthesize new quantum materials with novel properties and functionalities. Quantum chemistry, which uses quantum mechanics to study the behavior of atoms and molecules, is enabling the development of new catalysts, drugs, and functional materials, while quantum materials science is exploring the use of topological insulators, superconductors, and other exotic materials for applications such as quantum sensing, computing, and energy harvesting.

Finally, quantum research also engages with social and ethical considerations, as the development and deployment of quantum technologies raise new questions and challenges for privacy, security, and equity. Researchers are working with policymakers, ethicists, and social scientists to explore quantum technologies' societal implications and develop frameworks for responsible research and innovation in the quantum domain.

The Role of Philosophy

As quantum research delves deeper into uncharted territories and challenges our classical understanding of reality, it also poses profound philosophical questions about the foundations of physics and the nature of knowledge itself. These questions are not merely academic or abstract but have significant implications for understanding and interpreting the results of quantum experiments and theories.

One of the most fundamental philosophical questions in quantum mechanics is the nature of the wave function and its relationship to reality. Is the wave function a purely mathematical tool for calculating probabilities, or does it represent a real physical entity that evolves in space and time? Different interpretations of quantum mechanics, such as the Copenhagen interpretation, the many-worlds interpretation, and the pilot wave theory, offer different answers to this question and have significant implications for understanding the nature of measurement, causality, and determinism in the quantum world.

Another philosophical question is the role of the observer in quantum measurements and whether consciousness or human agency plays a special role in collapsing the wave function and determining the outcome of quantum experiments. While most physicists reject the idea of a special role for consciousness in quantum mechanics, the question remains a subject of ongoing debate and investigation.

Quantum research also engages with philosophical questions about the nature of space, time, and causality and whether these concepts need to be revised or reinterpreted in light of quantum phenomena such as entanglement and nonlocality. Some researchers have suggested that quantum mechanics may require a new framework for understanding causality and the flow of time, such as the retrocausal or atemporal interpretations of quantum mechanics.

Finally, quantum research also engages with epistemological questions about the nature of scientific knowledge and the limits of human understanding. As quantum mechanics challenges our classical intuitions and pushes the boundaries of what we can see and measure, it raises questions about the role of theory, experiment, and interpretation in the scientific process and the extent to which our cognitive and perceptual limitations constrain our knowledge of the quantum world.

7.5: Quantum Mechanics and Consciousness

Debates on Consciousness

One of the most fascinating and controversial areas of quantum research is the potential connection between quantum mechanics and consciousness. For decades, scientists and philosophers have debated whether the counterintuitive and strange features of quantum mechanics, such as entanglement, superposition and the measurement problem, may be related to the equally mysterious phenomenon of subjective experience and self-awareness.

At the heart of this debate is whether consciousness is a fundamental aspect of reality or an emergent property of complex physical systems such as the brain. Some researchers have suggested that quantum mechanics may provide a framework for understanding the nature of consciousness, and for explaining how subjective experience can arise from objective physical processes.

Others, however, have argued that the connection between quantum mechanics and consciousness is speculative and unscientific, and that there is no compelling evidence for a causal link between the two phenomena. Critics have pointed out that the brain is a warm, wet and noisy environment far from the ideal conditions for maintaining quantum coherence and entanglement, and that the time scales of neural processes are much slower than the time scales of quantum events.

Despite these criticisms, the debate over quantum consciousness is still a topic of intense interest and investigation among researchers from a wide range of fields, including physics, neuroscience, philosophy and psychology. Some of the most prominent theories and models of quantum consciousness include the Orch-OR theory proposed by physicist Roger Penrose and anesthesiologist Stuart Hameroff, which suggests that quantum processes in microtubules within neurons may give rise to conscious experience, and the quantum brain dynamics theory proposed by physicist Henry Stapp, which posits that the collapse of the wave function in the brain may be related to the flow of subjective experience.

Other researchers have explored the relationship between quantum mechanics and consciousness from a more philosophical and interpretive perspective, drawing on ideas from Eastern and Western traditions of thought. For example, some have suggested that the principles of quantum mechanics, such as nonlocality and the observer effect, may be consistent with certain aspects of Buddhist and Hindu philosophy, such as the concept of interconnectedness and the role of the mind in shaping reality.

Theoretical Proposals

One of the most well-known theoretical proposals for a quantum basis of consciousness is the Orch-OR theory, developed by physicist Roger Penrose and anesthesiologist Stuart Hameroff. According to this theory, consciousness arises from quantum processes within microtubules, tiny protein structures found within neurons in the brain.

Penrose and Hameroff suggest that these microtubules can sustain quantum coherence and entanglement over relatively long time scales and that they may be involved in the collapse of the wave function during the objective reduction (OR) process. They propose that this OR process, triggered by the accumulation of quantum gravitational effects, may give rise to moments of conscious experience and decision-making.

While the Orch-OR theory has attracted significant attention and debate within the scientific community, it remains very speculative and has been criticized on theoretical and experimental grounds. Critics have argued that the brain is too warm and noisy to sustain quantum coherence over the time scales required for the Orch-OR mechanism to operate and that there is no direct evidence for the involvement of microtubules in conscious processes.

Another theoretical proposal for a quantum basis of consciousness is the quantum brain dynamics (QBD) theory, developed by physicist Henry Stapp. According to QBD, the brain can be modeled as a quantum system that evolves according to the laws of quantum mechanics, and the collapse of the wave function during measurement may be related to the flow of subjective experience.

Stapp suggests that the brain may be able to maintain quantum coherence over relatively long distances and time scales and that the collapse of the wave function may be triggered by the interaction between the brain and its environment, including the sensory input from the body and the external world. He proposes that this collapse process may give rise to the unity and continuity of conscious experience and may be related to the binding problem in neuroscience.

While the QBD theory has also attracted significant interest and debate, it remains highly speculative and has been criticized on theoretical and experimental grounds. Critics have argued that the brain is too complex and heterogeneous to be modeled as a single quantum system and that the time scales of neural processes are much slower than the time scales of quantum events.

Scientific Perspectives

Despite the speculative nature of many quantum theories of consciousness, a growing body of scientific research is exploring the potential connections between quantum mechanics and the brain. This research spans a wide range of fields, from neuroscience and psychology to physics and computer science, and uses a variety of theoretical and experimental approaches to investigate the role of quantum processes in neural function and behavior.

One area of active research is the study of quantum effects in biological systems, such as photosynthesis, avian navigation, and the human sense of smell. These studies have shown that quantum coherence and entanglement can significantly improve the efficiency and sensitivity of biological processes and have raised questions about the potential role of quantum effects in the brain.

An exciting area of research is the advancement of quantum technologies for investigating the brain. These technologies include quantum sensors and imaging techniques that can analyze neural activity with a level of accuracy that was previously impossible. By utilizing the principles of quantum mechanics like superposition and entanglement, these technologies can transform our understanding of the brain and help us gain new insight into the relationship between neural activity and consciousness.

Two examples of quantum technologies used for studying the brain are magnetoencephalography (MEG) and optogenetics. MEG uses highly sensitive quantum sensors to detect the tiny magnetic fields produced by neural activity. It has been used to study various cognitive and perceptual processes, such as language, memory, attention, and emotion, and has provided new insights into the temporal and spatial dynamics of brain function.

On the other hand, optogenetics uses light-sensitive proteins and quantum dots to control and monitor neural activity with high precision and specificity. It has been used to study neural circuits involved in learning, memory, and behavior, providing new insights into the causal relationships between neural activity and cognitive function.

While these studies and technologies provide new insights into the brain and its relationship to consciousness, they still need to provide conclusive evidence for a causal link between quantum processes and subjective experience. Many scientists remain skeptical that consciousness is fundamentally quantum in nature and argue that classical physical processes can fully explain the brain at the level of neurons and synapses.

Philosophical Implications

The debate over quantum consciousness also has significant philosophical implications, as it touches on deep questions about the nature of reality, the relationship between mind and matter, and the limits of scientific knowledge. For centuries, these questions have been debated by philosophers and scientists, and have taken on new urgency and significance in light of the discoveries of quantum mechanics.

One of the key philosophical questions raised by quantum consciousness is the nature of the relationship between the subjective experience of consciousness and the objective physical world. Some researchers have suggested that consciousness may be a fundamental aspect of reality and that the physical world may be an emergent property of conscious experience rather than vice versa.

This view, known as idealism, has a long history in Western and Eastern philosophy and has been articulated by thinkers such as George Berkeley, Immanuel Kant, and the Indian philosopher Shankara. According to idealism, the physical world is a construct of the mind, and the ultimate nature of reality is mental or spiritual rather than material.

Another philosophical question quantum consciousness raises is the nature of free will and agency. Suppose consciousness is fundamentally quantum in nature and is subject to its laws, such as indeterminacy and nonlocality. In that case, this may affect our understanding of human freedom and responsibility.

Some researchers have suggested that the inherent randomness and unpredictability of quantum processes may provide a physical basis for free will and that the collapse of the wave function during the measurement process may be related to the experience of choice and decision-making. Others, however, have argued that quantum indeterminacy is insufficient for genuine free will and that the laws of quantum mechanics are still ultimately deterministic and constraining.

The philosophical implications of quantum consciousness are complex and multifaceted and touch on some of human thought's deepest and most enduring questions. While there is no consensus among philosophers or scientists on these issues, the debate over quantum consciousness has helped to stimulate new ways of thinking about the nature of mind, matter and reality. It has opened up new avenues for interdisciplinary research and dialogue.

7.6: Learning Resources and How to Further Your Quantum Journey

Recommended Books and Articles

For those eager to dive deeper into the world of quantum mechanics and continue their quantum journey beyond the pages of this book, there are many excellent resources available to help guide and inspire further exploration. Whether you are a student, a researcher, or simply a curious learner, there is no shortage of books, articles, and other materials to help you deepen your understanding and appreciation of this fascinating field.

One of the best places to start is with some of the classic textbooks and popular science books that have helped to shape the field of quantum mechanics and introduce its concepts and ideas to a wide audience. Some recommended titles include:

- "The Feynman Lectures on Physics, Volume III: Quantum Mechanics" by Richard Feynman, Robert Leighton, and Matthew Sands - This classic textbook, based on Feynman's legendary lectures at Caltech, provides a clear and engaging introduction to quantum mechanics principles and applications, with a focus on the physical and conceptual foundations of the theory.

- "Quantum Mechanics: The Theoretical Minimum" by Leonard Susskind and Art Friedman - This book provides a concise and accessible introduction to quantum mechanics' mathematical formalism, focusing on the key concepts and techniques needed to understand quantum systems' behavior.

- "What is Real? The Unfinished Quest for the Meaning of Quantum Physics" by Adam Becker - This popular science book provides a fascinating and accessible account of the history and philosophy of quantum mechanics, focusing on the ongoing debates and controversies surrounding the interpretation of the theory.

- "Quantum Computing Since Democritus" by Scott Aaronson - This book provides a wide-ranging and entertaining introduction to the field of quantum computing, with a focus on the fundamental concepts and algorithms that underlie this exciting new technology.

Here are other recommended titles for those wishing to delve into these topics in greater detail:

- "The Structure and Interpretation of Quantum Mechanics" by R. I. G. Hughes - This book provides a comprehensive and accessible introduction to the philosophical and interpretive issues surrounding quantum mechanics, focusing on the role of measurement and probability.

- "Quantum Reality: Beyond the New Physics" by Nick Herbert - This book explores the philosophical and metaphysical implications of quantum mechanics, focusing on the nature of reality, consciousness, and the relationship between mind and matter. It is a fascinating and engaging read.

- "Quantum Enigma: Physics Encounters Consciousness" by Bruce Rosenblum and Fred Kuttner - This book provides a clear and accessible introduction to the mystery of consciousness and its potential connection to quantum mechanics, with a focus on the measurement problem and the role of the observer.

- "Quantum Philosophy: Understanding and Interpreting Contemporary Science" by Roland Omnes - This book provides a comprehensive and rigorous introduction to the philosophical and interpretive issues surrounding quantum mechanics, with a focus on the role of logic and probability.

Online Courses and Educational Videos

In addition to books and articles, there are also many excellent online courses and educational videos available that can help you learn about quantum mechanics and related topics more interactively and engagingly. These resources often include lectures, demonstrations, and hands-on exercises that can help you develop a deeper understanding and intuition for the concepts and techniques of quantum mechanics.

Some recommended online courses and educational videos include:

- "Quantum Mechanics for Everyone" on edX - This free online course, offered by Georgetown University, provides a comprehensive and accessible introduction to the principles and applications of quantum mechanics, focusing on the theory's physical and conceptual foundations.

- "Quantum Computing for the Determined" on YouTube - This free video series, created by Michael Nielsen, provides a clear and engaging introduction to the field of quantum computing, with a focus on the fundamental concepts and algorithms that underlie this exciting new technology.

- "Quantum Mechanics" on Khan Academy - This free online course is a comprehensive and accessible introduction to the mathematical formalism of quantum mechanics, focusing on the key concepts and techniques needed to solve problems and understand the behavior of quantum systems.

- "Quantum Physics I" on MIT OpenCourseWare - This free online course, based on the first semester of MIT's undergraduate quantum physics sequence, provides a rigorous and comprehensive introduction to the principles and applications of quantum mechanics, with a focus on the Schrödinger equation, wave functions, and the quantum harmonic oscillator.

Here are other recommended titles for those wishing to delve into these topics in greater detail:

- "The Quantum World: Quantum Physics for Everyone" on Harvard Online Learning - This free online course, offered by Harvard University, provides a comprehensive and accessible introduction quantum mechanics principles and applications, with a focus on the philosophical and interpretive issues surrounding the theory.

- "Quantum Mechanics and Quantum Computation" on YouTube - This free video series, created by Umesh Vazirani, provides a clear and engaging introduction to the field of quantum mechanics and quantum computing, with a focus on the fundamental concepts and algorithms that underlie these exciting new technologies.

- "The Quantum Experiment That Broke Reality" on YouTube - This free video, created by the PBS Space Time channel, provides a fascinating and accessible introduction to the double-slit experiment and its implications for our understanding of reality, with a focus on the role of the observer and the measurement problem.

- "The Interpretations of Quantum Mechanics" on Coursera - This free online course, offered by the University of Maryland, provides a comprehensive and rigorous introduction to the philosophical and interpretive issues surrounding quantum mechanics. It focuses on the different interpretations of the theory and their implications for our understanding of reality.

Research Journals and Conferences

For those interested in staying up-to-date with the latest developments and discoveries in quantum mechanics and related fields, there are also many excellent research journals and conferences available that can help you stay informed with the cutting-edge of the field. These resources often include original research papers, review articles, and presentations from leading experts and researchers in the field. They can provide a valuable window into the current state of the art and the future directions of quantum research.

Some recommended research journals and conferences include:

- "Physical Review Letters" - This prestigious journal, published by the American Physical Society, features original research papers and rapid communications in all areas of physics, including quantum mechanics, quantum computing, and quantum information science.
- "Nature Physics" - This leading journal, published by Nature Research, features original research papers and review articles in all areas of physics, including quantum mechanics, quantum materials, and quantum technologies.
- "Quantum Information Processing" - This specialized journal, published by Springer, features original research papers and review articles in quantum information science and technology, including quantum computing, quantum communication, and quantum cryptography.
- "International Conference on Quantum Technologies" - This annual conference, organized by the Institute of Physics, gathers leading researchers and experts worldwide to discuss the latest developments and discoveries in quantum technologies, including quantum computing, quantum communication, and quantum sensing.

Here are other recommended titles for those wishing to delve into these topics in greater detail:

- "Foundations of Physics" - This interdisciplinary journal, published by Springer, features original research papers and review articles in the foundations and philosophy of physics, including quantum mechanics, quantum gravity, and quantum information theory.
- "Studies in History and Philosophy of Science Part B: Studies in History and Philosophy of Modern Physics" - This specialized journal, published by Elsevier, features articles and original research papers in the history and philosophy of modern physics, including quantum mechanics, relativity, and cosmology.
- "International Workshop on Quantum Foundations" - This annual workshop, organized by the International Journal of Quantum Foundations, brings together leading experts and researchers to discuss the latest developments and discoveries in the foundations and interpretation of quantum mechanics.
- "Quantum Mechanics and Quantum Information: Physical, Philosophical and Logical Perspectives" - The Perimeter Institute for Theoretical Physics organizes this conference, which aims to bring together experts and leading researchers from the fields of physics, philosophy, and logic to discuss the latest discoveries and developments

in the interpretation and foundations of quantum mechanics and quantum information theory.

Communities and Forums

Finally, for those who are looking to connect with other quantum enthusiasts and learners, and to participate in debates and discussions about the latest developments and ideas in the field, there are also many excellent online communities and forums available that can provide a valuable platform for learning, sharing, and collaboration.

Some recommended communities and forums include:

- "Quantum Computing Stack Exchange" - This online community, hosted by Stack Exchange, provides a platform for asking and answering questions about quantum computing, quantum information science, and related topics, focusing on the field's technical and programming aspects.
- "Quantum Information and Quantum Computer Scientists" on Facebook - With thousands of members, this Facebook group is a platform for sharing articles, news and discussions about quantum information science and quantum computing, focusing on the latest research and developments.
- "r/QuantumComputing" on Reddit - This Reddit community provides a platform for sharing news, articles, and discussions about quantum computing and related topics, focusing on accessible and engaging content for a general audience.
- "Quantum Mechanics" on Physics Forums - This online community, hosted by Physics Forums, provides a platform for asking and answering questions about quantum mechanics and related topics, focusing on the theory's fundamental concepts and techniques.

Here are other recommended titles for those wishing to delve into these topics in greater detail:

- "Foundations of Quantum Mechanics" on FQXi Community - This online community is a platform for discussing the philosophical and conceptual issues surrounding quantum mechanics.
- "Quantum Foundations" on Google Groups - This Google Group, with over 1,000 members, provides a platform for sharing news, articles, and discussions about the foundations and interpretation of quantum mechanics, focusing on the latest research and developments.
- "Quantum Mechanics and Quantum Information" on Quora - With over 100,000 followers, this Quora topic provides a platform for asking

and answering questions about quantum mechanics and quantum information science, focusing on accessible and engaging content for a general audience.

- "Quantum Physics" on Physics Stack Exchange - This online community, hosted by Stack Exchange, provides a platform for asking and answering questions about quantum mechanics and related topics, with a focus on the fundamental concepts and techniques of the theory, as well as the philosophical and interpretive issues surrounding the field.

Conclusion

The world of quantum mechanics is a vast and endlessly fascinating landscape of ideas, concepts, and applications that continue to challenge and inspire us today. From the theory's fundamental principles and philosophical implications to the cutting-edge technologies and research frontiers shaping the field's future, there are countless opportunities for learning, discovery, and exploration in this exciting and rapidly evolving area of science.

The resources and recommendations provided in this chapter, including textbooks, online courses, research journals, and community forums, can help guide and support you on your quantum journey, whether you are a student, researcher, or curious learner.

Studying quantum mechanics is not just an intellectual pursuit but also a deeply human one. By grappling with the profound implications of this theory, we expand our scientific knowledge and our understanding of ourselves and our place in the cosmos.

Approach quantum mechanics with curiosity, humility, and wonder. Appreciate the intricacy and allure of the quantum world, and let it inspire you to ask profound questions about reality, the limits of knowledge, and the opportunities of scientific exploration.

Remember that you are part of a vibrant and welcoming quantum community. By connecting with others, sharing ideas, and collaborating on projects, you can help advance the field and shape the future of this remarkable scientific endeavor.

Quantum Mechanics Demystified
A Practical Guide

Chapter 8: Quantum Mechanics Demystified - A Practical Guide

In this final chapter, I'll provide a step-by-step guide to solving quantum problems, explore case studies of quantum technologies in action, introduce quantum simulation software for hands-on learning, develop intuition for quantum phenomena through visualization, bridge the gap between theory and practice, and offer guidance for those seeking to pursue advanced study and careers in quantum physics.

So, let's put on our thinking caps and embark on a practical journey through the quantum realm, where we will demystify complex concepts, develop problem-solving skills, and unlock the potential of quantum mechanics to transform our world!

8.1: Step-by-Step Guide to Solving Quantum Problems

Problem-Solving Strategies

Quantum mechanics, with its abstract concepts and mathematical complexity, can often seem daunting to students and practitioners alike. However, with the right problem-solving strategies and a systematic approach, even the most challenging quantum problems can be tackled confidently and clearly.

The first step in solving any quantum problem is to carefully read and understand the problem statement and identify the system's key components, such as the potential energy function, boundary conditions and initial state. This step is crucial, as it sets the stage for the rest of the problem-solving process and ensures a clear grasp of the physics involved.

Next, it's essential to identify the relevant principles and equations that govern the quantum system's behavior. Depending on the problem, this may involve recalling fundamental concepts such as the Schrödinger equation, operator formalism, or perturbation theory. A strong grasp of these principles is essential for progressing toward a solution.

Once the relevant principles have been identified, the next step is to apply the appropriate mathematical techniques to solve the problem. This may involve techniques such as variable separation, Fourier analysis, or matrix diagonalization. It's essential to be comfortable with these mathematical tools and practice applying them to various quantum systems.

Breaking down complex quantum problems into smaller, more manageable parts is an effective strategy. For instance, if you aim to determine a complex quantum system's wave functions and energy levels, you may begin the process by solving the energy levels of a related, more straightforward system. For example, you can solve the energy levels of a particle in a box or a harmonic oscillator to achieve your end goal. By understanding the behavior of these simpler systems, you can gain insight into the more complex problem and develop a roadmap for solving it.

Another useful strategy is to use symmetry and conservation laws whenever possible. Many quantum systems possess inherent symmetries, such as rotational or translational invariance, which can greatly simplify problem-solving. By identifying these symmetries and using them to simplify the equations of motion, you can reduce the complexity of the problem and make it more tractable.

Worked Examples

To illustrate these problem-solving strategies in action, let's consider a few worked examples of common quantum problems.

Example 1: Particle in a Box

Think about a particle of mass m confined to a one-dimensional box of length L. The potential energy is zero inside the box and infinite outside. Find the energy levels and wave functions of the system.

Step 1: Identify the key components of the system.

- The particle has mass m and is confined to a one-dimensional box of length L.

- The potential energy inside the box is zero and infinite outside.

Step 2: Identify the relevant principles and equations.

1 The Schrödinger equation governs the behavior of the particle.

2 The boundary conditions require the wave function to be zero at the box's walls.

Step 3: Apply the appropriate mathematical techniques.

- Solve the Schrödinger equation using the method of separation of variables.

- Apply the boundary conditions to determine the allowed energy levels and wave functions.

Step 4: Interpret the results.

- The energy levels are given by $E_n = n^2 h^2 / (8mL^2)$, where n is a positive integer.

- The wave functions are given by $\psi_n(x) = \sqrt{2/L} \sin(n\pi x/L)$, where n is a positive integer.

Example 2: Harmonic Oscillator

Consider a particle of mass m in a one-dimensional harmonic oscillator potential $V(x) = (1/2) k x^2$, where k is the spring constant. Find the energy levels and wave functions of the system.

Step 1: Identify the key components of the system.

- The particle has mass m and is subject to a harmonic oscillator potential $V(x) = (1/2) k x^2$.

Step 2: Identify the relevant principles and equations.

- The Schrödinger equation governs the behavior of the particle.

- The harmonic oscillator Hamiltonian is given by $H = p^2/(2m) + (1/2) k x^2$.

Step 3: Apply the appropriate mathematical techniques.

- Solve the Schrödinger equation using the separation method of variables and the creation and annihilation operator formalism.

- Determine the allowed energy levels and wave functions using the properties of the harmonic oscillator.

Step 4: Interpret the results.

- The energy levels are given by $E_n = (n + 1/2) \hbar\omega$, where n is a non-negative integer and $\omega = \sqrt{k/m}$ is the angular frequency of the oscillator.

- The wave functions are given by $\psi_n(x) = (1/\sqrt{2^n n!}) (m\omega/\pi\hbar)^{1/4} e^{-m\omega x^2/(2\hbar)} H_n(\sqrt{m\omega/\hbar}\, x)$, where $H_n(x)$ are the Hermite polynomials.

These examples illustrate the power of a systematic approach to solving quantum problems. By breaking down the problem into its key components, identifying the relevant principles and equations, applying the appropriate mathematical techniques, and interpreting the results, you gain a deep understanding of the behavior of quantum systems and develop the skills needed to tackle more complex problems.

It's important to note that these examples are only the tip of the iceberg regarding the diversity of quantum problems you may encounter. Other common issues include the hydrogen atom, the quantum harmonic oscillator in two or three dimensions, the quantum tunneling problem, and the quantum spin problem, to name just a few. By practicing these problem-solving strategies on a wide range of quantum systems, you can develop a robust toolkit for tackling any quantum problem that comes your way.

Common Pitfalls

While developing problem-solving skills is essential for mastering quantum mechanics, it's equally important to be aware of the common pitfalls and misconceptions that can trip up even the most diligent students.

One common pitfall is the temptation to apply classical intuition to quantum systems. While classical mechanics provides a useful starting point for comprehending the behavior of macroscopic objects, it often fails to capture the strange and counterintuitive phenomena that emerge at the quantum scale. For example, the idea of a particle having a definite position and momentum at all times is a cornerstone of classical mechanics. Still, it breaks down in the quantum world due to the Heisenberg uncertainty principle.

Another pitfall is the tendency to get bogged down in the mathematical formalism of quantum mechanics at the expense of physical understanding. While a solid grasp of linear algebra, differential equations, and complex analysis is essential for solving quantum problems, it's equally important to develop a physical intuition for the behavior of quantum systems. This means taking the time to visualize wave functions, understand the physical meaning of operators and eigenvalues, and connect the mathematical results to real-world phenomena.

A third pitfall is failing to check your results for consistency and physical reasonableness. It's all too easy to make a small mistake in the algebra or to forget a factor of 2π somewhere along the way, leading to results that are off by orders of magnitude. By taking the time to double-check your work, compare your results to known solutions or limiting cases, and think critically about the physical implications of your answers, you can catch errors early and develop a deeper understanding of the problem at hand.

Finally, a common pitfall is the tendency to get discouraged or frustrated when confronted with a particularly challenging problem. Quantum mechanics is a complex subject that even the brightest minds have found difficult to understand. However, the key to mastering it lies in being patient, persistent and willing to learn from your mistakes. You can achieve this by breaking down each problem into smaller parts, asking for help from instructors or peers when necessary and celebrating your successes. With practice, you can develop the confidence and resilience needed to tackle any quantum challenge that comes your way.

Tips for Success

To help you avoid these common pitfalls and succeed in your study of quantum mechanics, here are a few tips to keep in mind:

1. ***Start with a solid foundation in classical mechanics and linear algebra.*** While quantum mechanics is a fundamentally different theory from classical mechanics, many of the same mathematical techniques and physical principles carry over. You'll be well-equipped to tackle the more advanced quantum mechanics concepts by mastering the basics of classical mechanics and linear algebra.

2. ***Practice, practice, practice.*** Like any skill, solving quantum problems requires practice and repetition to develop proficiency. Make a habit of working through problems regularly, whether from textbook exercises, online resources, or past exams. The more you practice, the more comfortable and confident you'll become with the problem-solving process.

3. ***Seek out multiple perspectives.*** Quantum mechanics can be approached from many different angles, and there's no one "right" way to solve a problem. By seeking out multiple perspectives, whether from different textbooks, online resources, or instructors, you can better understand the subject and develop a broader toolkit for problem-solving.

4. ***Collaborate with others.*** Quantum mechanics benefits greatly from collaboration and discussion. Working with classmates or joining a study group allows you to share ideas, troubleshoot problems, and learn from each other's strengths and weaknesses. One of the best ways to solidify your understanding is to explain a concept to someone else.

5. ***Embrace the uncertainty.*** One of quantum mechanics most difficult aspects is the theory's inherent uncertainty and probabilistic nature. Rather than trying to eliminate this uncertainty or find deterministic answers, embrace it as a fundamental feature of the quantum world. By learning to think about probabilities and distributions, you can develop a more instinctual understanding of quantum phenomena and become more comfortable with uncertainty.

6. ***Take advantage of visualization tools.*** Quantum mechanics is a highly abstract and mathematical subject, but that doesn't mean it has to be devoid of visual aids. Many excellent visualization tools and online simulations can help you develop intuition for quantum phenomena, from the double-slit experiment to the hydrogen atom. By exploring these visualizations and playing with the parameters, you can gain a more tangible understanding of the behavior of quantum systems.

7. ***Stay curious and keep learning.*** Quantum mechanics is a vast and constantly evolving field, with discoveries and applications always emerging. By remaining curious and learning about the latest developments in quantum physics, you can stay engaged and motivated

in your studies. Attend seminars, read popular science articles, and follow the work of leading researchers in the field.

Problem-Solving Checklist

Use this problem-solving guide to keep track of your problem-solving process and avoid common pitfalls. This checklist breaks down the problem-solving process into six key steps:

1. Read and understand the problem statement.
2. Identify the key components of the quantum system.
3. Determine the relevant principles and equations.
4. Apply the appropriate mathematical techniques.
5. Check your results for consistency and physical reasonableness.
6. Interpret your results and conclude.

By following this checklist and checking off each step as you go, you can ensure you approach each problem systematically and thoroughly. Note your observations at each step, to help you track your progress and identify areas where you may need additional practice or support.

The problem-solving checklist can also be valuable for collaboration and peer review. By sharing your completed checklist with classmates or instructors, you can get feedback on your problem-solving process and identify areas where you may have missed a key step or made an error. This collaborative learning is essential for developing a deep understanding of quantum mechanics and building a strong foundation for future study and research.

Whether you're a student just starting on your quantum journey, a researcher pushing the boundaries of quantum technology, or simply a curious learner wanting to explore the mysteries of the universe, the skills and strategies outlined in this sub-chapter will serve you well.

So grab your pencil and paper, fire up your favorite quantum simulation software, and dive into the exciting world of quantum problem-solving!

8.2: Case Studies in Modern Technologies

Quantum Mechanics in Technology

The principles of quantum mechanics are not just abstract theoretical concepts; they are the foundation of many of the most transformative technologies of the modern age. From the semiconductors that power our computers and smartphones to the lasers that enable high-speed communication and precision manufacturing, quantum mechanics is at the heart of the devices and systems that shape our world.

In this sub-chapter, we will explore a series of case studies that illustrate the crucial role of quantum mechanics in modern technologies. By examining how quantum principles are translated into practical applications, we will gain a deeper appreciation for the power and potential of this remarkable theory.

Case Study 1: Semiconductors and the Quantum Basis of Computing

The foundation of modern computing is the semiconductor, a material designed to control the flow of electrons precisely. The movement of electrons in semiconductors is governed by the principles of quantum mechanics, which dictate how electrons can occupy specific energy levels and traverse through a crystal lattice.

Semiconductor technology relies on a crucial quantum phenomenon called the band gap. This refers to the difference in energy between a material's valence and conduction bands. By controlling the composition and structure of semiconductors, engineers can adjust the band gap to create materials with specific electronic properties, such as conductivity or light emission.

Another crucial quantum effect in semiconductors is tunneling, which allows electrons to pass through thin layers of insulating material that would be impenetrable in the classical world. Quantum tunneling is the basis for many modern electronic devices, such as flash memory, which stores information by trapping electrons in a floating gate electrode.

Engineers have created various semiconductor devices by understanding and harnessing these quantum effects, from the simple diode to the complex microprocessor. These devices form the building blocks of modern computing, enabling the rapid processing and storage of vast amounts of information.

The impact of semiconductor technology on modern society can hardly be overstated. From the personal computer revolution of the 1980s to the smartphone boom of the 2000s, semiconductors have transformed how we live, work and communicate. They have enabled new industries, such as e-commerce and social media, and have revolutionized fields as diverse as medicine, transportation, and entertainment.

As we look ahead to the future, progress and innovation in semiconductor technology will play a crucial role in addressing some of humanity's biggest challenges. These challenges range from tackling climate change to eradicating diseases. By expanding the limits of what is achievable through quantum-based computing, we can develop novel tools and technologies that will aid us in creating a more equitable, prosperous, and sustainable world.

Case Study 2: Lasers and the Quantum Nature of Light

Another technology that owes its existence to the principles of quantum mechanics is the laser, which has become an indispensable tool in fields ranging from telecommunications to medicine to manufacturing.

At its core, a laser is a device that generates a highly coherent and focused beam of light through a process called stimulated emission. This process, first described by Albert Einstein in 1917, relies on the quantum nature of light and how electrons can absorb and emit photons.

In a laser, electrons in a special material called a gain medium are excited to a higher energy level by an external energy source, such as another laser or electrical current. When an incoming photon of the right wavelength interacts with an excited electron, it can stimulate the electron to emit a second photon with the same phase, frequency, and direction as the first. This process can be repeated often, leading to a cascading effect that generates a highly coherent and intense beam of light.

The quantum nature of the laser allows it to generate light with remarkable properties, such as high intensity, narrow bandwidth, and long coherence length. These properties have enabled a wide range of applications, from fiber-optic communication networks that carry vast amounts of data worldwide to precision manufacturing systems that can cut and shape materials with nanoscale accuracy.

One of the most exciting applications of laser technology is quantum computing, where lasers manipulate and control individual atoms or ions that serve as quantum bits or qubits. By carefully controlling laser light's frequency, phase, and intensity, researchers can create complex quantum states and perform quantum operations with high fidelity.

Another promising application of lasers is quantum cryptography, where the quantum nature of light is used to create secure communication channels immune to eavesdropping. By encoding information in the polarization or phase of individual photons and using the principles of quantum entanglement and superposition, researchers can create unbreakable encryption keys used to protect sensitive data.

From Theory to Application

The journey from quantum theory to practical application can be complex. It often requires a deep understanding of the underlying physics and a willingness to think creatively and take risks.

One key challenge in translating quantum principles into technology is the issue of scale. While quantum effects are readily observable at the level of individual atoms and molecules, they can be much harder to control and manipulate at the macroscopic scale of everyday devices and systems.

To overcome this challenge, engineers and scientists must develop new materials, fabrication techniques and control systems to harness the power of quantum mechanics in a reliable and scalable way. Practical application often requires a highly interdisciplinary approach, bringing together experts from physics, chemistry, materials science and electrical engineering.

Another challenge in developing quantum technologies is decoherence, which refers to how quantum systems can lose their delicate quantum properties when interacting with the environment. Decoherence is a major obstacle to building large-scale quantum computers and other quantum devices, as it can cause errors and information loss.

Researchers are developing new techniques for isolating and protecting quantum systems from external noise and interference to combat decoherence. These include using specialized materials, such as superconductors and topological insulators, advanced error correction, and fault-tolerant design principles.

Challenges in Application

Despite quantum technologies' many successes, significant challenges must be overcome to realize their full potential. One of the biggest challenges is scalability, which refers to the difficulty of building quantum devices and systems that can operate at large scales reliably and efficiently.

For example, while small-scale quantum computers have been demonstrated in the lab, building a large-scale quantum computer with millions of qubits remains a daunting challenge. This is partly due to the difficulty of maintaining the coherence and entanglement of large qubits and the need for advanced error correction and fault-tolerant design principles.

Another challenge is reproducibility and standardization. Unlike classical technologies, which can be mass-produced using standardized manufacturing processes, quantum technologies often require highly specialized and customized fabrication techniques, making reliability and consistency across different devices and systems difficult.

One effective method of tackling the difficulties associated with quantum devices and systems is to build modular and scalable architectures. This approach involves breaking down complex quantum systems into smaller, more manageable components. Researchers can create building blocks that can be combined and scaled up more reliably and efficiently, thus making the process easier and more streamlined.

Another approach is to use hybrid quantum-classical systems, which combine the strengths of quantum and classical computing to solve complex problems. Researchers can create more robust and scalable quantum systems that tackle real-world applications by using classical computers to control and optimize quantum devices.

Future Applications

Despite the remaining challenges, the future of quantum technologies looks incredibly promising. As we deepen our understanding of quantum mechanics and develop new materials, fabrication techniques and control systems, we can anticipate the emergence of a wide range of exciting new applications.

One area that holds particular promise is quantum sensing and metrology, which use quantum systems exquisite sensitivity to detect and measure physical quantities with unprecedented accuracy and precision. These technologies could have applications in medical imaging, drug discovery, navigation and exploration.

One promising area of research is quantum simulation. It involves using quantum computers to simulate complex quantum systems that classical computers cannot handle. This could provide scientists with a better understanding of the behavior of materials, chemicals, and biological systems. It may even lead to the development of new drugs, materials and technologies.

Perhaps the most exciting application of quantum technologies is in quantum communication and cryptography, which could enable secure communication networks that are fundamentally unbreakable. Using the principles of quantum entanglement and superposition, researchers could create communication channels immune to eavesdropping and hacking and enable new forms of secure communication and transactions.

Another promising application is quantum machine learning, which combines the power of quantum computing with the techniques of machine learning and artificial intelligence. Processing vast amounts of data through quantum computers allows researchers to create new algorithms and models to solve problems in finance, logistics and healthcare.

As we look to the future, quantum technologies will play an increasingly important role in shaping our world. From computing and communication to sensing and simulation, quantum mechanics' principles will continue to drive innovation and discovery in countless fields and industries.

However, realizing the potential of these technologies will require a sustained effort by researchers, engineers, and policymakers to invest in developing new quantum materials, devices, and systems. It will also require a commitment to education and outreach to ensure that the next generation of scientists and innovators is equipped with the knowledge and skills needed to advance the boundaries of what is possible.

Ultimately, the story of quantum technologies is one of human ingenuity, curiosity, and perseverance. It is a story of the incredible power of the human mind to unravel the mysteries of the universe and to harness those mysteries for the betterment of society.

As we continue on this journey of discovery and innovation, we must remain committed to the values of openness, collaboration, and transparency that have driven scientific progress for centuries. We must also remain mindful of these technologies' societal and ethical implications and work to ensure they are developed and deployed responsibly and equitably.

By working together and remaining true to the spirit of scientific inquiry and exploration, we can create a future in which quantum technologies help solve some of humanity's greatest challenges, from climate change and disease to poverty and inequality. This future is within our reach if we have the courage and vision to pursue it.

8.3: Quantum Simulation Software

Simulation Software Overview

As the field of quantum mechanics continues to advance, so too do the tools and technologies that researchers and students use to explore and understand its principles. One of the most powerful of these tools is quantum simulation software, which allows users to model and simulate the behavior of quantum systems on classical computers.

Quantum simulation software comes in many different forms and flavors, from simple educational tools designed for beginners to advanced research-grade platforms used by experts in the field. Some of the most popular and widely used quantum simulation platforms include:

• QuTiP: An open-source Python framework for simulating and analyzing quantum systems, focusing on quantum optics and information processing.

• Qiskit: An open-source quantum computing platform developed by IBM, which includes a wide range of tools for building and simulating quantum circuits and access to real quantum hardware through the IBM Quantum Experience.

• Cirq: An open-source quantum computing library developed by Google, which provides a high-level interface for building and simulating quantum circuits and integration with Google's quantum hardware.

• Microsoft Quantum Development Kit: A comprehensive platform for developing and simulating quantum algorithms and applications, which includes the Q# programming language, a quantum simulator, and integration with Microsoft's Azure Quantum service.

These platforms and many others offer a wide range of features and capabilities, from simple visualizations and interactive simulations to advanced programming interfaces and access to real quantum hardware.

One of the key advantages of quantum simulation software is that it allows researchers and students to explore and experiment with quantum systems in a way that would be impossible or impractical with real quantum hardware. By simulating the behavior of quantum systems on classical computers, users can gain valuable insights into the underlying principles of quantum mechanics and develop and test new algorithms and applications without the need for expensive and complex quantum hardware.

Another advantage of quantum simulation software is that it can help bridge the gap between theory and experiment. By providing a virtual laboratory for exploring quantum phenomena, simulation software can help researchers develop and refine their theories and identify new avenues for experimental investigation. This can accelerate the pace of discovery and innovation in the field of quantum mechanics and help bring new technologies and applications to market more quickly.

Getting Started with Quantum Simulations

For those new to quantum simulation software, getting started can seem daunting. However, with a little guidance and practice, anyone can begin to explore and experiment with these powerful tools.

One of the first steps in starting with quantum simulations is to choose a platform appropriate for your expertise and experience. For beginners, educational tools like QuVis or Quantum Playground can be a great place to start, as they provide simple, interactive simulations to help build intuition and understanding of basic quantum concepts.

For those with more advanced skills and experience, research-grade platforms like QuTiP, Qiskit, or Cirq can provide a more powerful and flexible environment for exploring and simulating quantum systems. These platforms typically require some programming and quantum mechanics knowledge but offer advanced users a wide range of features and capabilities.

Once you have chosen a platform, the next step is to familiarize yourself with its interface and capabilities. Most quantum simulation platforms provide extensive documentation and tutorials to help users get started, as well as community forums and support resources for troubleshooting and learning from others in the field.

One key skill to develop when working with quantum simulation software is the ability to translate quantum problems and algorithms into code. This requires a solid understanding of the underlying mathematical principles of quantum mechanics and proficiency in programming languages like Python, C++, or Q#.

To develop these skills, starting with simple examples and tutorials and gradually working up to more complex and challenging problems can be helpful. Many quantum simulation platforms provide a wide range of pre-built examples and demos that can be a starting point for learning and experimentation.

Another important skill to develop is the ability to visualize and interpret the results of quantum simulations. Quantum systems can be highly complex and counterintuitive, and it can be challenging to make sense of the output of a simulation without the right tools and techniques. Most quantum simulation platforms provide a range of visualization and analysis tools to help users understand and interpret their results, from simple plots and animations to advanced statistical analysis and machine learning techniques.

Practical Exercises

To help you get started with quantum simulation software, here are a few practical exercises that you can try on your own:

Exercise 1. Simulating a Qubit

One of the simplest quantum systems to simulate is a single qubit, which can be in a superposition of two states (0 and 1) and manipulated using quantum gates and operations.

Using a quantum simulation platform of your choice, try to create a simple simulation of a single qubit and explore how different quantum gates (such as the Hadamard gate or the Pauli-X gate) affect its state.

Exercise 2: Simulating Quantum Entanglement

Quantum entanglement is one of the most fascinating and counterintuitive phenomena in quantum mechanics and is a key resource for many quantum technologies and applications.

Using a quantum simulation platform, try to create a simulation of a simple entangled state (such as the Bell state) and explore how measurements on one qubit can affect the state of the other.

Exercise 3: Simulating Quantum Algorithms

Quantum algorithms are special sequences of quantum gates and operations that can solve certain computational problems faster than classical algorithms.

Using a quantum simulation platform, try to implement a simple quantum algorithm (such as the Deutsch-Jozsa algorithm or the Grover search algorithm) and compare its performance to a classical algorithm for the same problem.

These exercises are just a starting point, and there are many more advanced and challenging problems that you can explore using quantum simulation software. Some other ideas for practical exercises include:

- Simulating quantum error correction codes protects quantum information from noise and errors.

- Simulating quantum cryptography protocols, which use the principles of quantum mechanics to enable secure communication and information processing.

- Simulating quantum machine learning algorithms, which use quantum computers to speed up certain machine learning tasks, such as classification and clustering.

By working through these exercises and exploring the capabilities of quantum simulation software, you can gain a deeper understanding of quantum mechanics principles and applications and develop the skills and knowledge needed to contribute to this exciting and rapidly evolving field.

Understanding Limitations

While quantum simulation software is a powerful and valuable tool for exploring and understanding quantum mechanics, it is essential to recognize its limitations and approximations.

One of the main limitations of quantum simulation software is that it runs on classical computers, fundamentally different from quantum computers in terms of their architecture and capabilities. This means that quantum simulations on classical computers are inherently limited by the resources and performance of the underlying hardware and may not fully capture the complexity and scale of real quantum systems.

Another limitation of quantum simulation software is its reliance on approximations and simplifications to make the simulation tractable on classical computers. For example, many quantum simulation platforms use "trotterization" to break up a complex quantum system into smaller, more manageable pieces and simulate each separately. While this can effectively simulate large quantum systems, it can also introduce errors and inaccuracies into the simulation.

In addition to these technical limitations, quantum simulation software has conceptual limitations that are important to keep in mind. One of the most fundamental of these is that quantum systems are inherently probabilistic and non-deterministic, which means that the outcomes of quantum measurements and operations are not always predictable or repeatable.

This inherent randomness and uncertainty can make designing and interpreting quantum simulations challenging and limit the results' accuracy and reliability. It is important to recognize that quantum simulations are always an approximation of the true behavior of a quantum system and that there will always be some level of uncertainty and error in the results.

Another conceptual limitation of quantum simulation software is that it can sometimes obscure the underlying physical principles and intuitions behind quantum mechanics. By abstracting away the details of the quantum hardware and focusing on the mathematical formalism of quantum operations and measurements, simulation software sometimes makes it difficult to develop a deep understanding of the physical basis of quantum phenomena.

To mitigate these limitations, it is vital to use quantum simulation software with other tools and techniques, such as analytical calculations, experimental measurements, and physical intuition. By combining these different approaches and perspectives, researchers and students can gain a more complete and accurate understanding of the behavior and properties of quantum systems.

8.4: Developing Intuition for Quantum Phenomena through Visualization

Visualizing Quantum Concepts

One of the most significant challenges in learning and understanding quantum mechanics is the abstract and counterintuitive nature of many key concepts and phenomena. From wave-particle duality and quantum superposition to entanglement and the uncertainty principle, the behavior of quantum systems can be difficult to visualize and comprehend using classical intuition alone.

To help overcome this challenge, researchers and educators have developed a wide range of visualization tools and techniques that can help to build intuition and understanding of quantum phenomena. These tools range from simple static images and diagrams to interactive simulations and immersive virtual reality experiences. They can be used to explore and explain a wide range of quantum concepts and applications.

One of the most basic and widely used visualization techniques in quantum mechanics is using wave functions and probability distributions to represent the state of a quantum system. By plotting the wave function of a quantum particle or system, researchers and students can gain insight into the system's spatial distribution and temporal evolution and visualize how different quantum operations and measurements affect the system's state.

Another powerful visualization technique in quantum mechanics is using Bloch spheres to represent the state of a qubit. A Bloch sphere is a three-dimensional representation of a two-level quantum system, where the north and south poles represent the two basis states of the qubit (typically denoted as $|0\rangle$ and $|1\rangle$), and any point on the surface of the sphere represents a superposition of these two states.

By using Bloch spheres to visualize the state of a qubit, researchers and students can gain insight into the geometric and topological properties of quantum states. They can explore how different quantum gates and operations affect the qubit's state. Bloch spheres can also be used to visualize the effects of quantum noise and decoherence on the state of a qubit and to design and optimize quantum error correction codes and fault-tolerant quantum circuits.

Interactive Visualizations

While static images and diagrams can be useful for building intuition and understanding of quantum concepts, interactive visualizations can take this understanding to the next level by allowing users to explore and manipulate quantum systems in real-time.

One of the most powerful examples of interactive quantum visualization is quantum circuit simulators, which allow users to design and simulate quantum circuits using a graphical user interface. These simulators typically include a wide range of pre-built quantum gates and operations, as well as tools for measuring and analyzing the quantum system's state.

By using quantum circuit simulators, researchers and students can explore the behavior of quantum algorithms and protocols and gain hands-on experience with the design and optimization of quantum circuits. These simulators can also be used to study the effects of quantum noise and errors on the performance of quantum circuits and to develop and test new quantum error correction schemes.

Interactive quantum visualization can be achieved through virtual reality (VR) and augmented reality (AR) technologies, which create immersive and engaging quantum learning experiences. By wearing VR headsets and using controllers, users can explore and interact with three-dimensional representations of quantum systems and phenomena. This allows them to better understand the spatial and geometric properties of quantum states and operations.

For example, researchers at the University of Bristol have developed a VR experience called "The Quantum Playground," which allows users to explore and manipulate the state of a single qubit using a virtual Bloch sphere. By using hand gestures and body movements, users can apply quantum gates and operations to the qubit, and can observe how these operations affect the state of the system in real-time.

Other examples of VR and AR quantum visualizations include the "Quantum Odyssey" experience developed by the Institute for Quantum Computing at the University of Waterloo, which takes users on a journey through the history and applications of quantum mechanics, and the "Quantum Sandbox" developed by the QuTech research center in the Netherlands, which allows users to build and simulate their own quantum circuits using a virtual reality interface.

The Role of Visualization in Learning

The use of visualization tools and techniques can play a crucial role in the learning and understanding of quantum mechanics, particularly for students and researchers who are new to the field.

Visualizations play a crucial role in understanding the abstract and counterintuitive concepts of quantum mechanics. They present a visual and intuitive representation of these concepts, which can make the learning process more engaging and effective. Additionally, visualizations help to highlight the connections and relationships between different quantum phenomena and applications, providing a more holistic and integrated view of the field.

Moreover, the use of interactive visualizations and simulations can provide students and researchers with hands-on experience and practice with quantum systems and algorithms, which can help to reinforce learning and build practical skills. By immersing oneself in a virtual environment and conducting experiments with quantum systems, individuals can acquire a more profound comprehension of the behavior and limitations of these systems. This, in turn, can help them develop intuition and expertise that can be effectively utilized in the real-world applications of quantum technologies.

The effectiveness of visualization in quantum learning has been demonstrated in numerous studies and educational settings. For example, a study by researchers at the University of St Andrews found that students who used interactive simulations and visualizations in a quantum mechanics course performed significantly better on assessments than those who did not, and reported higher levels of engagement and understanding of the material.

Another study by researchers at the University of Pittsburgh found that the use of visual analogies and metaphors can be particularly effective in helping students to understand and remember complex quantum concepts. By comparing quantum phenomena to more familiar and intuitive concepts, such as waves on a string or spinning tops, instructors can help students to build mental models and intuitions that can be applied to more abstract and mathematical aspects of quantum mechanics.

The use of visualization in quantum learning is not limited to the classroom or laboratory setting, but can also be used in public outreach and science communication efforts. By creating engaging and accessible visualizations of quantum concepts and technologies, researchers and educators can help to build public understanding and support for quantum research and development, and can inspire the next generation of quantum scientists and engineers.

Examples of Successful Visualizations

There are many examples of successful visualizations in quantum mechanics that have helped to advance the field and build understanding and intuition among students and researchers. Here are a few notable examples:

1. The "Quantum Corral" image, created by researchers at IBM's Almaden Research Center in 1993, uses scanning tunneling microscopy to visualize the wave function of electrons confined to a circular "corral" on a copper surface. The image, which shows the electron waves forming a standing wave pattern within the corral, helped to demonstrate the wave-particle duality of electrons and the potential for using quantum effects in nanotechnology and quantum computing.

2. The "Quantum Lego" project, developed by researchers at the Institute for Quantum Computing at the University of Waterloo, uses 3D-printed

models and interactive exhibits to visualize and explain quantum concepts and technologies to the general public. The project includes a set of building blocks that represent different quantum gates and operations, and allows users to build and explore their own quantum circuits in a hands-on and engaging way.

3. The "Quantum Game with Photons" project, developed by researchers at the University of Bristol and the University of Oxford, uses a web-based game to teach players about the basics of quantum mechanics and quantum cryptography. The game, which is based on the BB84 quantum key distribution protocol, challenges players to use quantum states of light to securely transmit and receive messages, while learning about the principles of quantum superposition, measurement, and entanglement.

4. The "Quantum Odyssey" virtual reality experience, developed by the Institute for Quantum Computing at the University of Waterloo, takes users on a journey through the history and applications of quantum mechanics, from the early experiments of the 20th century to the cutting-edge technologies of today. The experience, which uses a combination of 3D animation, narration, and interactive elements, helps users to visualize and understand the key concepts and phenomena of quantum mechanics, such as the double-slit experiment, the Schrodinger equation, and quantum entanglement.

5. The "Quantum Sandbox" virtual reality experience, developed by researchers at QuTech in the Netherlands, allows users to build and simulate their own quantum circuits using a virtual reality interface. The experience includes a set of pre-built quantum gates and operations, as well as tools for measuring and analyzing the state of the quantum system, and allows users to explore the behavior and limitations of quantum circuits in a hands-on and immersive way.

These examples demonstrate the power and potential of visualization in quantum mechanics, and highlight the diversity of tools and techniques used to build understanding and intuition in this complex and fascinating field. As quantum technologies continue to advance and evolve, the role of visualization in quantum learning and discovery is likely to become even more important and impactful.

Visualizations play a crucial role in developing intuition and comprehension of quantum phenomena, especially for beginners in the field, like students and researchers. They provide an intuitive and graphical representation of abstract and counterintuitive concepts, which can help in bridging the gap between theory and understanding. This can make the learning process more engaging and effective. In conclusion, visualizations are a valuable tool that can aid in comprehending quantum concepts.

The use of interactive visualizations and simulations, such as quantum circuit simulators and virtual reality experiences, can provide hands-on experience and practice with quantum systems and algorithms, which can help to reinforce learning and build practical skills. Moreover, the use of visualization in public outreach and science communication efforts can help to build public understanding and support for quantum research and development, and can inspire the next generation of quantum scientists and engineers.

As the field of quantum mechanics evolves and advances, visualization plays a crucial role in learning and discovery. Keeping current with the latest developments and technologies in quantum visualization and experimenting with these powerful tools can put you at the forefront of this rapidly evolving field, allowing you to leverage its potential to the fullest.

8.5: Quantum Mechanics in Action

Bridging Theory and Practice

The study of quantum mechanics is not only a theoretical pursuit but also a significant foundation for developing a wide range of groundbreaking technologies and applications. The principles of quantum mechanics are being utilized to resolve real-world problems and create new opportunities for innovation and discovery, ranging from quantum computing and cryptography to quantum sensing and imaging.

To bridge the gap between the theory and practical applications of quantum mechanics, it is crucial to develop a deep understanding of the underlying principles and mathematical formalism of the field, as well as a practical knowledge of the tools and techniques used to design, simulate, and implement quantum systems and algorithms.

One of the key ways to gain this practical knowledge is through hands-on experience with quantum hardware and software platforms. Many universities, research institutions, and companies now offer access to quantum computing and simulation resources, either through cloud-based services or on-site facilities, which allow students and researchers to experiment with real quantum systems and develop their skills and expertise.

Another important aspect of bridging theory and practice in quantum mechanics is the development of interdisciplinary collaborations and partnerships between researchers, engineers, and industry partners. Quantum technologies and applications often require expertise from a wide range of fields, including physics, mathematics, computer science, and electrical engineering, and the most successful projects are often those that bring together teams with diverse backgrounds and skillsets.

For example, the development of practical quantum computing systems requires close collaboration between theoretical physicists who understand the underlying principles of quantum information and computation, experimental physicists who can design and build the physical hardware, computer scientists who can develop the software and algorithms, and electrical engineers who can design the control and readout electronics. By collaborating and sharing resources across disciplines, teams can speed up innovation and launch quantum technologies more efficiently.

Real-World Problem Solving

One of the most exciting aspects of quantum mechanics in action is the potential for using quantum technologies and algorithms to solve real-world problems that are intractable or inefficient using classical methods.

Quantum computing has immense potential to transform the fields of drug discovery and materials science. It can help in simulating and optimizing complex molecular systems that classical computers cannot handle. By using quantum algorithms like the variational quantum eigensolver (VQE) and the quantum approximate optimization algorithm (QAOA), scientists can explore large design spaces of chemicals and materials. They can discover new compounds and materials with desired properties such as targeted drug activity or high-temperature superconductivity.

Another example of real-world problem solving with quantum mechanics is the use of quantum sensing and imaging technologies to detect and diagnose diseases at the cellular and molecular level. By using techniques such as quantum-enhanced magnetic resonance imaging (MRI) and quantum-enhanced positron emission tomography (PET), researchers can visualize and analyze biological systems with unprecedented sensitivity and resolution, enabling earlier and more accurate diagnosis and treatment of diseases such as cancer and Alzheimer's.

Quantum cryptography is another area where quantum mechanics is being used to solve real-world problems in information security and privacy. By using the principles of quantum entanglement and superposition, researchers can develop cryptographic protocols that are fundamentally unbreakable, even by future quantum computers. These protocols, such as quantum key distribution (QKD) and quantum-secured blockchain, have the potential to revolutionize fields such as finance, healthcare, and national security by providing a level of security that is impossible to achieve with classical methods.

Quantum sensing and metrology is another area where quantum mechanics is being used to solve real-world problems in fields such as navigation, geology, and environmental monitoring. By using techniques such as quantum-enhanced gravimetry and magnetometry, researchers can detect and measure physical quantities with unprecedented precision and sensitivity, enabling new applications such as underground resource exploration, earthquake detection, and climate monitoring.

Emerging Technologies

As the field of quantum mechanics continues to evolve and advance, new technologies and applications are emerging that have the potential to transform a wide range of industries and sectors.

One of the most exciting emerging technologies in quantum mechanics is quantum machine learning, which combines the power of quantum computing with the techniques of machine learning and artificial intelligence. By using quantum algorithms for tasks such as data classification, pattern recognition, and optimization, researchers can develop new machine learning models that can learn from and analyze vast amounts of data more efficiently than classical methods, with applications in fields such as finance, healthcare, and transportation.

Another emerging technology in quantum mechanics is quantum simulation, which uses quantum computers to simulate and analyze complex quantum systems that are beyond the reach of classical computers. By using quantum simulation to study systems such as high-temperature superconductors, topological insulators, and quantum chemistry, researchers can obtain new insights into the fundamental properties of matter and develop new materials and devices with novel properties and functionalities.

Quantum communication is another emerging technology that has the potential to revolutionize the way we transmit and process information. By using techniques such as quantum teleportation and quantum repeaters, researchers can develop long-distance quantum communication networks that can transmit quantum information securely and reliably over large distances, with applications in fields such as finance, defense, and scientific research.

Quantum sensing and imaging is another emerging technology that has the potential to transform fields such as healthcare, environmental monitoring, and materials science. By using techniques such as quantum-enhanced microscopy and spectroscopy, researchers can image and analyze biological and chemical systems with atomic-scale resolution and sensitivity, enabling discoveries and subsequent applications in fields such as drug discovery, disease diagnosis, and materials characterization.

The Future of Quantum Applications

As the field of quantum mechanics continues to evolve and mature, the future of quantum applications looks increasingly bright and transformative. With the rapid progress being made in quantum hardware, software, and algorithm development, there is a likelihood we will see a wide range of new and exciting quantum technologies and applications emerge in the coming years and decades.

One of the key areas where quantum applications are likely to have a major impact is in the field of artificial intelligence and machine learning. By combining the power of quantum computing with the techniques of deep learning and reinforcement learning, researchers may be able to develop new AI systems that can learn and adapt more efficiently and effectively than classical systems, with applications in fields such as robotics, autonomous vehicles, and personalized medicine.

Another area where quantum applications are likely to have a major impact is in the field of materials science and engineering. By using quantum simulation and optimization techniques to design and discover new materials with novel properties and functionalities, researchers may be able to develop new technologies such as high-temperature superconductors, topological insulators, and quantum metamaterials, with applications in fields such as energy, electronics, and sensing.

Quantum cryptography and communication are also likely to play an increasingly important role in the future of information security and privacy. As classical cryptographic methods become increasingly vulnerable to attacks by quantum computers, the development of quantum-secure communication protocols and networks will become essential for protecting sensitive information and transactions in finance, healthcare and national security.

Quantum sensing and metrology will likely impact greatly on a wide range of fields in the future, from fundamental physics and cosmology to environmental monitoring and resource exploration. By using quantum techniques to measure physical quantities with unprecedented precision and sensitivity, researchers may be able to detect new particles and forces, study the early universe and dark matter, and monitor the health and sustainability of our planet and its resources.

As we look to the future of quantum applications, it is clear that the field of quantum mechanics will continue to be a driving force for innovation and discovery across a wide range of disciplines and industries. By developing new technologies and applications that harness the power of quantum mechanics, we can create a sustainable, secure and prosperous future.

8.6: Preparing for Advanced Study in Quantum Physics

Advanced Topics in Quantum Mechanics

For those who have mastered the basics of quantum mechanics and are eager to explore more advanced topics and applications, there is a wide range of exciting areas to explore. These topics build upon the foundational principles and techniques covered in earlier chapters and require a deeper understanding of the mathematical and conceptual frameworks of quantum mechanics.

Quantum mechanics' advanced topics include quantum field theory, which provides a comprehensive framework to explain the behavior of particles and fields in relativistic quantum systems. Quantum field theory is essential to understanding the Standard Model of particle physics, which explains the fundamental particles and forces of nature, and to develop new theories of quantum gravity and cosmology.

Another advanced topic in quantum mechanics is quantum information theory, which studies the processing, transmission, and storage of information using quantum systems. Quantum information theory is the foundation for the development of quantum computing, quantum cryptography, and quantum communication, and has important applications in fields such as computer science, mathematics, and engineering.

Quantum many-body theory is another advanced topic in quantum mechanics that deals with the behavior of systems with large numbers of interacting particles, such as condensed matter systems, quantum fluids, and quantum gases. Quantum many-body theory is essential for understanding the emergent properties and phase transitions of complex quantum systems, and has important applications in fields such as materials science, condensed matter physics, and quantum chemistry.

Quantum optics is another advanced topic in quantum mechanics that studies the interaction of light and matter at the quantum level, and is fundamental for developing quantum technologies such as quantum sensors, quantum imaging and quantum communication. Quantum optics is also important for understanding the fundamental properties of light and the nature of quantum entanglement, and has important applications in fields such as atomic physics, quantum information processing, and quantum metrology.

Educational Pathways

To prepare for advanced study in quantum physics, it is crucial to follow a structured educational pathway that builds upon the foundational knowledge and skills acquired in earlier stages of learning. This pathway typically involves a combination of coursework, research experience, and independent study, and may vary depending on the specific area of quantum physics that one wishes to pursue.

At the undergraduate level, students interested in quantum physics should focus on building a strong foundation in mathematics, including differential equations, linear algebra and complex analysis, as well as in classical physics, including mechanics, electromagnetism, and thermodynamics. They should also take introductory courses in quantum mechanics and modern physics, which will provide a basic understanding of the principles and techniques of quantum theory.

At the graduate level, students should focus on more advanced coursework in quantum mechanics, quantum field theory, quantum information theory, and other specialized topics, depending on their research interests. They should also gain research experience by working with faculty members on projects in quantum physics, either through independent study, research assistantships, or internships.

For those interested in pursuing a career in quantum physics research, it is important to consider graduate programs that have strong research groups in the specific areas of interest, as well as access to state-of-the-art experimental and computational facilities. Some of the top graduate programs in quantum physics include those at MIT, Caltech, Stanford, UC Berkeley and Harvard in the United States, as well as international programs at institutions such as the University of Oxford, the University of Toronto and the Max Planck Institute for Quantum Optics.

For those interested in pursuing a career in quantum physics education or outreach, it is important to consider programs that have a strong emphasis on teaching and communication skills, as well as opportunities for public engagement and outreach. Some examples of such programs include the Physics Education Research group at the University of Colorado Boulder, the Center for Engaged Learning at Elon University, and the Quantum Outreach and Education program at the University of Waterloo.

Research Opportunities

One of the most exciting aspects of preparing for advanced study in quantum physics is the opportunity to engage in cutting-edge research and contribute to advancing the field. Many research opportunities are available to students and early-career researchers in quantum physics, ranging from theoretical and computational projects to experimental and applied work.

At the undergraduate level, students can gain research experience by working with faculty members on projects in quantum physics, either through independent study, research assistantships, or summer research programs. These projects can range from theoretical investigations of quantum algorithms and protocols to experimental studies of quantum materials and devices.

At the graduate level, students can pursue more advanced research projects as part of their thesis or dissertation work, either independently or in collaboration with other students and faculty members. These projects can involve the development of new quantum technologies and applications, the study of fundamental questions in quantum physics, or the exploration of interdisciplinary connections between quantum physics and other fields such as computer science, mathematics, and engineering.

There are also many research opportunities available to early-career researchers in quantum physics, such as postdoctoral fellowships, research scientist positions, and faculty positions at universities and research institutions. These positions often involve leading independent research projects, mentoring students and junior researchers, and collaborating with other researchers and industry partners.

Some examples of leading research institutions and centers in quantum physics include the Institute for Quantum Computing at the University of Waterloo, the Centre for Quantum Technologies at the National University of Singapore, the Joint Quantum Institute at the University of Maryland, and the quantum research groups at companies such as Google, IBM, and Microsoft. These institutions and centers offer a wide range of research opportunities in areas such as quantum computing, quantum communication, quantum sensing, and quantum materials, and provide access to state-of-the-art experimental and computational facilities.

Building a Career in Quantum Sciences

The field of quantum physics is rapidly expanding and diversifying, presenting numerous intriguing career prospects for individuals with advanced training and proficiency in the field. These opportunities are available across various industries and fields, including academia, government and the private sector. They offer a remarkable chance to make meaningful and substantial contributions to the progress of science and technology.

For individuals wanting to follow a career in academia, the typical path would involve obtaining a PhD in quantum physics or a related field, followed by gaining research experience as a postdoctoral researcher, and eventually securing a faculty position at a university or research institution. Pursuing an academic career in quantum physics presents the opportunity to conduct independent research, guide and mentor students and junior researchers, as well as teach at undergraduate and graduate levels.

For those interested in pursuing a career in government or national laboratories, there are many opportunities available at institutions such as the National Institute of Standards and Technology (NIST), the National Science Foundation (NSF) and the Department of Energy (DOE). These positions often involve conducting research in areas of national priority, such as quantum computing, quantum communication, and quantum sensing, and collaborating with academic and industry partners.

For those interested in pursuing a career in industry, there are many opportunities available at companies and startups working on the development and commercialization of quantum technologies and applications. These positions often involve working on research and development projects, product design and engineering, and business development and strategy.

Some examples of companies and startups working in the field of quantum physics include Google, IBM, Microsoft, IonQ, Rigetti, and D-Wave Systems. These companies are working on the development of quantum computing hardware and software, quantum communication and cryptography systems, and quantum sensing and imaging technologies, and offer varied career opportunities for those with expertise in quantum physics and related fields.

To build a successful career in quantum sciences, it is important to develop a strong foundation in the principles and techniques of quantum mechanics, as well as skills in areas such as programming, data analysis, and scientific communication. It is also important to stay up-to-date with the latest developments and trends in the field, and to build a network of colleagues and collaborators through conferences, workshops, and online communities.

Conclusion

In conclusion, preparing for advanced study in quantum physics is an exciting and rewarding journey that offers the opportunity to explore the frontiers of science and technology, and to make significant contributions to the advancement of the field. By following a structured educational pathway, engaging in cutting-edge research, and building a strong foundation in the principles and techniques of quantum mechanics, students and early-career researchers can position themselves for success in varied career opportunities in academia, government and industry.

The field of quantum physics is expanding and advancing rapidly. Consequently, the demand for well-informed and skilled professionals in this field is expected to increase in the near future. By maintaining a curious and creative mindset, and also by committing to lifelong learning, those who pursue advanced education in quantum physics can make a significant difference in the world. They can also contribute to shaping the future of science and technology.

A Quantum Leap Forward

As we conclude our journey through the captivating world of quantum mechanics, it's important to reflect on the ground we've covered, from the basic principles of wave-particle duality and quantum states to the mind-bending concepts of entanglement and quantum computing. Understanding quantum mechanics allows us to appreciate the fundamental workings of the universe and gain insights into the nature of reality.

The impact of quantum mechanics extends far beyond the laboratory, underpinning many technological marvels that shape our modern world. A conceptual understanding of these ideas offers its own rewards, cultivating a deeper appreciation for the beauty and elegance of the natural world.

As you close this book, carry forward the spirit of curiosity and exploration. The journey of discovery in quantum mechanics never truly ends, and there is always more to learn. Apply the principles of quantum mechanics in your own life, both in understanding the world around you and in driving innovation in your chosen field.

While we've covered a significant amount of ground, quantum mechanics is a vast and rapidly evolving field. Stay curious, stay updated with the latest developments, and never stop asking questions.

Thank you for your enthusiasm and dedication throughout this exploration. The beauty of quantum mechanics lies in its ability to inspire curiosity, wonder and a deep appreciation for the mysteries that surround us.

With an open mind, a curious heart and a commitment to lifelong learning, there is no limit to what we can discover in this endlessly fascinating field.

Made in United States
Troutdale, OR
09/30/2024

23267897R00100